Using
Computers
in the
**Behavioral
Sciences**

Using Computers in the Behavioral Sciences

Paul C. Cozby

California State University, Fullerton

Mayfield Publishing Company

Sponsoring editor: Franklin C. Graham
Manuscript editor: Marie Enders
Managing editor: Pat Herbst
Art director: Nancy Sears
Interior and cover designer: Paul Quin
Illustrator: Naomi Schiff
Production manager: Cathy Willkie
Production coordinator: Jane B. Radosevic
Compositor: Kachina Typesetting, Inc.
Printer and binder: George Banta Company

Library of Congress Catalog Card Number: 83-062821
International Standard Book Number: 0-87484-714-1

Manufactured in the United States of America
10 9 8 7 6 5 4 3 2 1

Mayfield Publishing Company
285 Hamilton Avenue
Palo Alto, California 94301

Contents

Appendix
Computer Programming   159

Preface

Using Computers in the Behavioral Sciences is an introduction to computer methods for students majoring in fields such as psychology, sociology, anthropology, political science, education, communications, and speech. In all of these fields of study, the use of computers is becoming increasingly necessary as both a research tool and a job skill. It is thus becoming necessary for educators in the social sciences to provide instruction in computer use to students in their respective fields.

This book will be useful for instructors in a variety of settings. For example:

- As a supplementary text for instructors of research methods or statistics who wish to include a goal of computer literacy as part of the course.
- As the main text or one of several texts in a social science course designed to teach computer uses to majors. The text is sufficiently brief to allow instructors to provide additional material to tailor the course to their unique needs and purposes.
- As an entry-level required reading for students in advanced degree programs to ensure that all students in the program have minimal computer knowledge despite the diverse backgrounds of the students.

For most purposes, chapters may be assigned in any order, and chapters or portions of chapters may be omitted to suit the needs of individual instructors.

Acknowledgments

I would like to thank several individuals for their help and support in the preparation of this book. Art Graesser and Jeanne King provided useful feedback on much of the manuscript. Annette Gilbert "word-processed" with amazing speed and accuracy and never complained about messy notes in the margin. Dan Kee, along with Art Graesser, almost daily provided me with interesting discussions about computers in the social sciences. Frank Graham at Mayfield was supportive and provided gentle prods to keep me working. Monica Turner, my department secretary, covered up for me as I neglected some of my duties. Hiro Yasuda supplied a great deal of technical assistance. Finally, I am most grateful for the valuable suggestions and reactions of the following reviewers: Bernardo J. Carducci of Indiana University Southeast, Helen J. Crawford of the University of Wyoming, Fred Leavitt of California State University at Hayward, Gil Osgood of the University of Oregon, Donald E. Pannen of the University of Puget Sound, and James M. Weyant of the University of San Diego.

P.C.C.

Introduction to Computing

Computers have become a fact of life in our society. Almost daily we read about or see on television some new application of computer technology. For many years large industries and government have used computer capabilities for a variety of purposes. Now computers are readily available to all of us, and the range of their uses has expanded rapidly. Computers are being used in education, for the statistical analysis of data in all of the social and behavioral sciences, in psychological testing and clinical practice, in biofeedback techniques for stress reduction, in preparation of manuscripts and books, and in scientific instrumentation and experimental manipulations.

These developments have made it almost a necessity for students of the social and behavioral sciences to acquire computer literacy. Computer literacy demands that you become familiar with how computers operate and how they are programmed and, most important, that you *use* a computer. The purpose of this book is to help students become computer literates. As you proceed, try to remember that a computer is only a machine that obeys commands given to it by you, the computer user. Have fun!

What Is a Computer?

A computer is simply a machine that consists of a large number of electronic circuits. But it is a very special machine, one that is designed to process information. The terms *information* and *data* refer to such things as the results of an experiment or a survey, a letter or manuscript, and a list of names and addresses—in general, numbers, words, instructions, or whatever someone might have that a computer can understand and work with to make a specific task easier or more efficient. A computer will receive information given to it by a user, process that information in whatever way the user desires, and then provide the user with the processed output. Most important, the computer can process large amounts of information with great speed and accuracy.

Special Purpose Computers
Two types of computers are manufactured, special purpose computers and general purpose computers. *Special purpose computers* are designed to perform a very limited number of specific functions. For example, many new automobiles are designed with a computer that monitors the engine and makes adjustments in the carburetor and other engine components to maximize fuel economy or reduce pollu-

tion emission. A pocket calculator is also a special purpose computer as is the computer that people can program to cook their food in a microwave oven. Finally, many of us have purchased one of the home video game computers designed to play game programs by means of buttons and joystick controls. More and more of the products we use contain a special purpose computer that usually makes the operation of the product more sophisticated or efficient (although one might question whether most of us really need many of these devices, such as a programmable microwave oven when a simple timer would suffice).

General Purpose Computers

General purpose computers are designed to perform any function the user desires. Here the user writes or purchases computer programs each of which performs a different function. Many people have purchased their own general purpose computer—a microcomputer such as the Apple™, TRS-80™, Atari™, or any number of other "personal" or "home" computers. Such computers may be used to do statistical calculations, bookkeeping, income tax preparation, word processing; to teach math, play games, or do almost anything a user might want to do on a computer.

General purpose computers are typically categorized according to their size. Larger computers can process more information with greater speed than can smaller computers.

Main-frame computers are the very large machines that are needed by large businesses and government and for much scientific research. Tasks such as processing census data, keeping track of airline schedules and reservations, and maintaining student records in a large university demand the use of a main-frame computer. Main-frames are large and fast and can be used simultaneously by many people.

Minicomputers were first developed in the 1960s. Because these computers were smaller and less expensive than other existing computers, computer technology and applications were made directly available to businesses, scientists, and others who previously could not have afforded or justified a large computer. Minicomputers are not able to process as much information or handle as many users simultaneously as the larger main-frame computers.

The 1970s brought us the *microcomputer*. Advancements in engineering technology allowed an entire computer to be built in a very small space—on a single circuit board—and very inexpensively. Computers were made available to many more people; in fact, this technology has advanced to the point at which some manufacturers are selling handheld computers and computers that sell for less than $100. What has made this possible is the *microprocessor*, an entire computer brain built as a single small silicon chip. Microprocessors have also made possible many of the special purpose computers that are becoming so prevalent.

These distinctions are useful, but new technology is blurring the dividing lines between main-frames, minis, and micros. For example, a new type of microprocessor enables small microcomputers to have more information processing capability than the minicomputers of just a few years ago. New "super-mini" computers are more powerful than many main-frames. Perhaps most interesting is the ability to communicate between computers. New networking systems allow a user with a small microcomputer at a desk to communicate with other users with their microcomputers, to use the capabilities of a mini-computer located somewhere else, or to become a user of a large main-frame computer that might even be located in another state.

An Apple II microcomputer system. The keyboard is the input device and the video monitor serves as the output device. The small printer can be used for obtaining printed output. Two disk drives are used for auxiliary memory. (Courtesy of Apple Computer, Inc.)

This CRT-type computer terminal is connected to a PDP 11/70 computer. Similar terminals placed in other locations allow many people to use the computer simultaneously. (Courtesy of Digital Equipment Corporation)

Advances in computer technology occur at an amazing rate, so all of us can expect to continue to see an expansion of the capabilities of computers.

Information Processing on a Computer

There are three components of information processing on a computer: input, computer storage and processing, and output. The computer user must first input information into the computer; the computer's task is to store that information and do whatever processing the user desires; finally, the computer must provide output back to the user.

As a simple example, suppose a user wishes the computer to add a series of numbers. The user would need to input a program that instructs the computer to add numbers, input the numbers themselves, and then input instructions to execute the program. The computer would store the program and the numbers and when told to do so would add them together. Finally, the computer would have to output the results to the user.

Because most of us only see the input and output components of this process, these will be described first. The actual mechanics of the computer will then be described.

Input Devices
There are a number of ways to communicate information to a computer. Most common is a typewriter-like keyboard. Information is typed

A computer operator working with the IBM System/370 computer. The computer is connected to a printer and several disk drives that serve as information storage devices. (Photo courtesy of IBM Corporation)

on a line, and when the "carriage return" key is depressed, the information on the line is sent to the computer. The keyboard is usually connected to a television screen (cathode-ray tube or CRT) or a printer so the user can see what is being typed. The CRT or the printer will also serve as the output device. The entire unit is called a computer terminal (see Box 1–1).

Before the keyboard became widely used, punched cards were the most common input device. Computer cards are placed in a keypunching machine. The machine punches a different pattern of holes for each letter of the alphabet, number, or special symbol. The cards are then placed in a card reader that deciphers the pattern of holes and sends the information to the computer. Punched cards are not considered desirable for most applications today because they don't allow the fast interaction with the computer that a computer terminal allows.

Disk storage and tape storage are other means of input. With disk storage, information is coded on a round disk. Different materials may be used for the disk—large data processing facilities use a hard disk whereas small computer systems use small floppy disks made of flexible vinyl. A disk drive machine has the capability to "write" information on the disk and also "read" information on the disk and input the information to the computer. With tape storage, information is coded on magnetic tape. A tape drive can be run forward or backward to write new information on the tape or read existing

Computer cards are punched and then placed in a card reader.

Box 1–1   Communicating with the Computer

The most common way of communicating with a computer is via a computer terminal with a keyboard and either a CRT or printer. If you are working on a microcomputer, it is likely that you will be the sole user of the computer. In many settings, however, you will be sharing the computer with others. Sometimes terminals are hard-wired directly to the computer; at other times you may have to communicate via a telephone connection. With telephones, a phone number is dialed and connection made. The receiver is placed in a *modem* (or acoustic coupler), a device designed to translate computer signals to signals that can be sent over a phone line. The user's modem translates signals from the terminal and sends these to a modem connected to the computer. The computer's modem makes another translation so the signal can be received by the computer. This process works in reverse when information is sent from the computer back to the terminal.

To use the computer, you will have to learn how to "log-on" to the computer. Your instructor will show you the log-on procedure for your machine. On a typical system, the computer will request that you provide both an account number and some sort of password. This is necessary for system security and to prevent computer use by unauthorized persons. (Because of the danger of people stealing or altering the information in computers, computer scientists are continually making their security systems more sophisticated.)

Information to be sent to the computer is typed on a line on the screen. A cursor shows you where you are typing. While typing on a line, you can correct mistakes with a key that will backspace and then retype the information. When the return key is depressed, information on the line is sent to the computer.

There are also some special keys on the terminal that aren't on a standard typewriter, such as "control" and "escape." Because each terminal and computer has its own unique characteristics, your instructor will have to show you which of these keys are on your system and how they are used.

information into the computer. Tape and disk storage will be discussed further below.

Although keyboards, punched cards, disks, and tapes are the most common input devices, others are available, usually for special uses. Forms that require a number 2 pencil are designed to be read by a special device (an optical character reader or OCR) that reads the location of the pencil marks; usually the device codes this information on tape that is later input to the computer for processing. Similarly, the numbers that appear on the bottom of your bank checks are read by a machine, and then the information is input to the computer to maintain your checking account. Joysticks and buttons that appear on Pac-Man and other video games are also input devices that tell the computer where to move the Pac-Man, monster, or star destroyer. You have also probably seen scanners in grocery stores that use a laser device to read the Universal Product Code printed on grocery labels. Another development is the "mouse" used on the Lisa Computer built by Apple Corp. The mouse is a device used to move a cursor anywhere on the screen. Because the screen shows all commands the user needs, the user doesn't have to type commands on the keyboard. Finally, voice input is becoming a reality. What was once a science fiction scenario of a user "talking" to a computer may become an increasingly common mode of input. Currently, the most widespread use of voice input is with computers and computer-controlled aids designed for handicapped individuals.

Output Devices

After the computer processes information, the user requires that the results be made available on an output device. Again, several methods are available. As noted above, television screens (CRT's) and printers are the most common. The obvious disadvantage of a CRT is that the output is not permanent. Thus most users of a CRT computer terminal will have access to a printer that can be used when hard copy is needed. Less expensive low-speed printers are frequently attached directly to the terminal or shared by a group of users. However, when there is a vast amount of output (such as printing the names of every person in each class at a large university, printing the payroll checks of everyone at a large corporation, or even performing a complex statistical analysis), expensive high-speed printers that are connected to large computers are needed. These printers are able to print several thousand lines per minute. In addition to speed of output, printer users must be concerned with the quality of the printed copy, whether the printer can make multiple copies through carbons, and whether graphics are to be printed. The choice of printers depends upon the requirements of the user.

Disk storage and tape storage are also used as output devices. Computers are frequently asked to revise and update information and then output that information to disk or tape storage where it can be accessed later. A good example is word processing. The advantage of word processing is that a user can input something like a letter and store the letter on a disk or tape (as output). Later the letter may be input from the disk or tape and revised in whatever way is desired; then the new version may be stored for printing or revising whenever necessary.

Finally, new output devices are being made possible with advanced technology. Music and voice synthesizers are the most noteworthy. A large proportion of young children today are receiving their first exposure to computers with devices such as the Speak and Spell™ manufactured by Texas Instruments.

Computer Processing

The computer itself is responsible for receiving the input, processing it, and directing the output to an output device. Obviously, the computer is the crucial component even though the user doesn't actually see the inner workings of the computer.[1] In this section, computer hardware will be described.

The computer has two components, main memory and the central processing unit (CPU). The small CPU used in microcomputers is called a microprocessor. Main memory is the information storage component, and the CPU is the "brain" of the computer that is responsible for all information processing. These components are manufactured in the form of silicon chips that contain a vast amount of electronic circuitry in a very small amount of space. Chips encased by a plastic covering are mounted on circuit boards that provide the necessary interconnections.

Main Memory

Information in a computer is stored in the form of binary digits called bits (*b*inary dig*it* = bit). A *bit* of information is simply a single on or off signal. Within the computer, this is an electronic circuit that is either open or closed to the flow of electricity. This binary system means that at the most fundamental level, information is coded very

The circuit board for the Apple II computer. Most of the chips are used for main memory storage. One chip is the central processing unit (CPU). The slots on the far right are used to connect the computer to a printer and/or disk drive. (Courtesy of Apple Computer, Inc.)

simply yet very precisely—a bit can take on the value of either a 0 (zero) or a 1 (one), a "no" or a "yes."

A single bit cannot be used to code very complex information. In fact, a bit by itself can only be used to code two digits such as 0 or 1, or two letters of the alphabet such as *A* or *B* (e.g., the bit could store an *A* if the circuit is on and a *B* if the circuit is off). Obviously, not much can be done with a bit. In the computer, bits are grouped together as a unit called a *byte*. Several bits working together as a byte can store complex pieces of information because the byte can take on many on-off combinations of the individual bits. For example, a byte composed of three bits may take on eight different values ($2 \times 2 \times 2 = 8$):

0	0	0
0	0	1
0	1	0
0	1	1
1	0	0
1	0	1
1	1	0
1	1	1

Even this is limited, however, A 3-bit byte could not even distinguish among all 26 letters of the alphabet. A 4-bit byte could take on 16 combinations ($2 \times 2 \times 2 \times 2 = 16$), a 5-bit byte would have 32 combinations, and so on. Depending upon the particular computer, most machines will have bytes that contain 6 or more bits. The computer is engineered so that every letter, number, punctuation mark, and symbol (e.g., $+$, $-$, etc.) has its own unique pattern within the byte.

Microcomputers generally have a minimum of 4096 bytes. This is referred to as 4K—one K is 2^{10} bytes or 1024 bytes.[2] Thus a 4K computer has 4096 bytes of memory. A 64K machine has 65,536 bytes. The memory in large mini- and main-frame computers is indexed in terms of megabytes—a megabyte is about one million bytes. It is easy to see now why larger computers can handle more information; there are simply more bytes in main memory.

Each byte may be thought of as a memory address that identifies it. That is, each piece of data is stored in main memory in a unique location—a particular memory address—that can be accessed by the computer. The fact that every single piece of information exists in a unique location is extremely important if the user is going to be able to have the computer do things with the information. For example, if we store the number 10 in one memory address and the number 20 in another location, the computer can be directed to access the numbers, add them together, and then output the results to a CRT. Such functions are carried out by the central processing unit.

Central Processing Unit

The central processing unit (CPU) is often called the brain of the computer. It is responsible for doing all the work that we ask the computer to perform. The CPU has two components necessary for its operation: the arithmetic/logic unit and the control unit.

The *arithmetic/logic unit* carries out arithmetic operations such as addition, subtraction, multiplication, and division. It also performs logical operations to do things such as compare whether two numbers are equal to, greater than, or less than one another.

The *control unit* controls the execution of instructions to the computer. It is responsible for receiving instructions from the user, fetching information stored in main memory, directing the arithmetic/logic unit to do things to the information, and then directing the sending of output. The control unit does this over and over again with great speed and accuracy. As an example, read the following description of the Apple II™ Plus processor taken from the owner's manual:

> This is the brain of your Apple. It is a Synertek/MOS Technology 6502 microprocessor. In the Apple, it runs at a rate of 1,023,000 machine cycles per second and can do over five hundred thousand addition or subtraction operations in one second. It has an addressing range of 65,536 eight-bit bytes.*

Larger computers have processors that are much faster and can handle much more information. Still, the basic operation of the computer with information stored in main memory and processing of information carried out by the CPU is fundamental to all computers.

Auxiliary Memory

Earlier we referred to disk storage and tape storage devices. These are connected to the computer and serve as *auxiliary memory* or permanent memory. One reason why auxiliary memory is necessary is that the information stored in main memory is "volatile"—that is, the memory is not permanent because it ceases to exist when the computer is turned off or when new information must be input for performing some new task. Imagine that you just spent an hour typing into the computer a program that would enable you to play a game. You then play the game for ten minutes but have to leave for a class. You would like to play the game again after class or anytime you wanted. However, the entire program will be lost because it is currently in main memory. Fortunately, if the computer is connected to an auxiliary memory device, the program can be *saved* on disk or tape. Later the information can be *loaded* into main memory (that is, transferred from the disk or tape) and the game can be played again. Permanent memory storage allows a computer to be used for a large variety of purposes. Main memory is used only temporarily while performing a given task, thus freeing the user to use main memory storage over and over again for subsequent tasks.

An auxiliary memory device is also necessary because the amount of storage in main memory is finite and thus limited. For example, many home computers have 48K bytes of main memory. (Recall that 1K is 1024 bytes.) 48K translates into 49,152 bytes. Now suppose that

---

*Apple II Reference Manual. (1981). Copyright © 1981 by Apple Computer, Inc. Used by permission of Apple Computer, Inc., 20525 Mariani, Cupertino, CA 95014.

someone writing a book wishes to use the word processing capabilities of this computer. How much of the book could be stored in main memory? The answer is 49,152 characters (letters, numbers, symbols, punctuation marks, blank spaces). If there are 60 characters per line and 15 lines per page, then only about 32 pages can be accommodated in main memory. The ability of the computer to transfer information between main memory and permanent memory is extremely important for performing many computer tasks.

Both tape and disk storage have been mentioned. Each has certain advantages and disadvantages. Magnetic tape is inexpensive as are the tape drive devices used to communicate with the computer. Home computer users can use a simple tape cassette recorder, for example. The drawback is that tape is very slow. Because the information on the tape is stored sequentially, you must literally wind through the tape to find the information needed (much as someone with an audiotape must wind through several songs before getting to the desired selection).

In contrast, disks and disk drive devices are relatively expensive. Even disk drives for home computers cost several hundred dollars. However, information retrieval using disks is very fast. Information on the disk can be located and loaded into a computer very quickly. To continue the music analogy, consider how quickly you can change from one song to another on a record.

Thus the decision to use tape or disk is a matter of cost versus speed and convenience. For many computer applications, both will be used. For example, disk storage may be used for information that is currently being worked on, updated, or analyzed. Other information that is used infrequently or being held for possible use later on may be transferred to tape.

To conclude, it should be made clear that a single tape or disk is able to contain only a certain amount of information. However, the storage capability is infinite because new tapes or disks can be used. Further, information on a tape or disk can be erased, leaving room for new information. Finally, although tapes and disks are the most common auxiliary storage devices, new methods of storing information are being developed. One such device is the "magnetic bubble," in which the information is stored as magnetic fields on strips of specially made materials. Another new storage technology is an optical digital disk that is based on the same principles that underlie many of the home videodisc systems.

Software

All of the computer hardware devices enable the computer to operate. However, they are simply machines that are useful only when there is software to give instructions to the computer. *Software* refers to the computer programs that are written to make the computer perform a certain task. As noted earlier, software enables us to apply the capabilities of the computer to uses such as word processing, teaching, testing, performing statistical analyses, keeping records, and so on.

As a computer user, you might encounter a computer program in several ways. First, you could learn the skills of computer programming and write programs for your own use. More commonly, many individuals in business, industry, education, and government specify a needed use of the computer and then employ computer programmers to write the necessary programs. Even more frequently, computer users compare and eventually purchase commercially available programs. These programs are sold as tapes or disks and come with written instructions for their use. Finally, more and more books are being sold that contain written programs. If someone wishes to use a program described in the book, the person must type the program into the computer. This development has been useful for owners of home computers who don't have the skills to write sophisticated programs of their own.

For most of us, *all* work on a computer is done with software programs. If you are working with a large computer at your school or company, the programs have been made available for all users of the system or you can probably write your own programs. If you have a small microcomputer, you will probably purchase programs on tape or disk. Alternatively, virtually all microcomputers have available the programming necessary for you to write programs in BASIC language as soon as you turn on the machine (see Box 1–2).

Summary

Computers are machines designed to process information. Special purpose computers perform a limited number of functions and are generally found in various products (such as microwave ovens). General purpose computers can perform any number of functions. Main-frame computers are the largest general purpose computers. They can process vast amounts of information and may be used by many people simultaneously. Minicomputers are smaller than main

Box 1–2   RAM and ROM

Y ou may have heard representatives of computer companies talk about having a computer with, for example, 48K of RAM and then also talk about additional memory in ROM. RAM stands for random access memory and is essentially blank memory addresses (bytes) ready to store information. ROM is read only memory; ROM consists of memory circuits prewired into the machine that already contain computer programs that are automatically available whenever the computer is turned on. For example, on most microcomputers, the computer programming that allows you to use BASIC (a relatively simple programming "language") is stored in ROM.

Special purpose computers necessarily consist largely of ROM because they are sold to perform very specific functions. For example, consider the video games in your local arcade or pizza parlor. ROM holds all the programs that instruct the computer to do such things as display a sequence of events on the screen when no one is playing and also execute the sequence of events in the game according to game rules (e.g., the speed of the enemy blasters). RAM holds information on the player's score, the current location of, say, the Pac-Man, and so on. On many arcade machines, the initials of the highest scorers are displayed as well.

If you wish to experiment with RAM and ROM, go to a video arcade that has a game that displays the initials of high scorers. Those initials are stored in RAM. Now ask the owner to unplug the game. A blank screen appears. When the game is plugged in again, something *will* appear on the screen. What appears is actually the result of a computer's executing programs that are stored in ROM.

frames in terms of processing capability and number of users. Micro-computers are the small "personal" computers that are used by a single user.

Information may be input to a computer on a keyboard or with punched cards. Tapes and disks on which information has been previously stored are other input devices. Still other input devices include optical character readers.

After information is processed by the computer, the results must be output to the user. A CRT (cathode-ray tube) or a printer is the most common output device. Results may also be output to tape or disk.

The computer itself has two components: main memory and the central processing unit (CPU). Main memory holds the information that the computer is working on. The amount of information that the computer can store is indexed in terms of bytes. One K of memory is 1024 bytes. The CPU contains a unit that controls the operation of the computer. It finds information in main memory, executes instructions, and directs output to the user. The arithmetic/logic unit carries out arithmetic operations and logical operations necessary for the processing.

Auxiliary memory or permanent memory devices are necessary to store information permanently because information in main memory is volatile. Disks and tapes are the most common means of permanently storing information.

Software refers to the computer programs that instruct the computer how to operate to perform various tasks. Software for computer applications such as statistical analyses is either developed by the user or purchased from a commercial supplier.

RAM refers to random access memory—blank memory space that a user may use for data input and computer programs. ROM is read only memory. ROM consists of computer programs that are built into the computer and cannot be modified by the user.

Further Reading

Keeping up with developments in the field of computer technology is a challenging task. Magazines such as *Byte, High Technology, Popular Computing,* and *Psychology Today* regularly report on new developments.

Computers and computation (1971). San Francisco: W. H. Freeman.
Science (1982, February 12). (Issue on computers and electronics)

Films

A series of films entitled *Adventure of the Mind* is available from the Indiana University Audio-Visual Center. Individual titles are

"The Personal Touch" (#BSC–183)
"Hardware and Software" (#BSC–184)
"Speaking the Language" (#BSC–185)
"Data Processing, Control, Design" (#BSC–186)
"For Better or for Worse" (#BSC–187)
"Extending Your Reach" (#BSC–188)

Notes

1. One enjoyable way to increase your intuitive understanding of the computer is to read a book by Tracy Kidder entitled *The Soul of a New Machine* (Little, Brown, 1981). Kidder is a journalist who followed the course of the manufacture of a new computer from the initial idea to the final product.
2. 2^{10} refers to an exponent. Exponentiation is simply multiplying a number by itself a specified number of times. Thus, $2^2 = 2 \times 2 = 4$; $2^3 = 2 \times 2 \times 2 = 8$; 2^{10} multiplies 2 by itself ten times and equals 1024. In computer language, $1K = 2^{10}$ bytes or 1024 bytes.

2

Information
Input and Storage

...AND OF THOSE FIVE NEWSPAPERS YOU READ
EACH DAY, WHICH WOULD YOU SAY IS THE BEST?

This chapter will outline the basic procedures that are necessary to input, store, and change information in the computer. Information input and storage is required for most computer uses, including record keeping, statistical analysis, and computer programming.

The Concept of Files

Most information that is input to the computer is organized in the form of *files*. Imagine the labels on the file folders that might exist in an office file cabinet: 1982 sales, projected budget, speech to support group, results of customer survey. Each file contains information on a single topic. Computers must allow users to create files in much the same way that they might create files for their home or office.

A computer must be able to do more than simply create a file. To be truly useful, it must be able to

- Add to the contents of a file
- Change existing information in a file for correcting or updating
- Delete information from a file
- Move information in a file from one location to another

As an illustration, suppose I wanted to type a book chapter using a word processer. An increasingly important application of computers is word processing. Word processing software is now available for most computers. On a typical system, I would first create a file with a particular name. I would type material using whatever margin settings I select. I wouldn't have to worry about stopping at the end of each line; the computer automatically "wraps" from one line to the next. It also automatically paginates so there is no need for me to worry about stopping at the end of each page. I should also be able to automatically center and underline material. When I've finished typing, I can permanently store the material on a disk. I can then print the material if I like.

Later, I can add material to my book chapter file. These additions may be placed at the end of the file or inserted at any point in the file. I should also be able to delete parts of the chapter I don't like and move material from one part of the chapter to another. It should also be easy to make corrections if I made any typing errors. On some systems, the computer will be able to search the entire file and change the spelling of a word that was misspelled throughout the chapter. Many word processors have the capability of checking each word in the file against a "dictionary" of words and then printing a list of any misspelled words. Finally, I can have the revised version of the book chapter file replace the original version on the disk.

The general process of file use on a computer is diagrammed in Figure 2–1. In the remainder of this chapter, we will explore the various computer commands that must be used to create and work with files. Because every computer system uses different commands, only general descriptions will be used and space will be provided for you to fill in information about the command used on your computer.

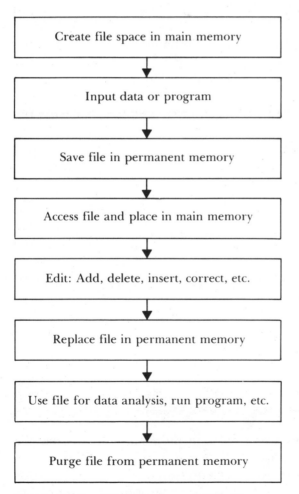

Figure 2–1   The Process of File Use

Creating a File

The first step of file use is to tell the computer that you wish to input a new file. On many computers, the command to do this is something like

NEW,filename

where the NEW command tells the computer to expect a new file and "filename" stands for whatever file name is given to the material. For example, if I create a file on the word processor for chapter 2 in a book, I might use the following command:

NEW,CHAP02

It is a good idea to have an informative file name—one that can easily be remembered.

On most systems, there are rules about file names in terms of length (e.g., no more than seven letters) and whether they can be more than one word. It is extremely important to remember that every file must have a unique name that is different from the name of every other file. Your computer may require that you name your file at this point or at the point when it is saved. Use the space below to note the method used on your system.

The command to create a new file on my system is _____

I name my file   ☐ when it is first created, or   ☐ when it is first saved (Check one.)

What are the limitations on file names?

_____

_____

_____

At this point, you are ready to input information. This may be scientific data, a computer program, a paper you have written, financial data, etc. On some systems, you can immediately begin typing your input. Other systems may require that you give another command to enter information. You will also have to use a command when you are through with your input.

I   ☐ simply begin typing, or   ☐ give another command (Check one.)

The command to enter information is _____

The command to stop entering information is _____

Saving a File

While you are creating a file, you are using the computer's main memory. The information will be lost if you leave the system or turn off the computer. Usually you will want to permanently save the file so that you can use it anytime you are using the computer. On many systems, the command to do this is simply

SAVE,filename

(Alternatively, the comma and file name can be deleted on some systems.) This command will cause the information to be saved on the computer's permanent memory system (e.g., disk or tape).

The command to save a new file on my system is _____

_____

_____

Other information about creating or saving a file on my system:

_____

_____

_____

Retrieving a File

Once a file has been permanently saved, you can then call for it to be either used by a program or modified by an editor. A command must be given to access the file from the auxiliary storage device and then load the information into main memory. Remember that you can only use or work with a file when the information in the file is in the computer's main memory. The file remains permanent on the tape or disk even when you are working with it in main memory. The terms *local* and *permanent* are frequently used when making this distinction. The local version of the file consists of the information currently in main memory, whereas the permanent version is the information stored in the auxiliary memory system. Thus, if you are typing in a new file, the information is *local* until you use the command to save the file. At that point, the file becomes *permanent.*

In order to make a permanent file local in main memory, a command to retrieve the file must be used. An example of such a command is

OLD,filename

where the term OLD tells the computer to find a file that has previously been saved (contrast this OLD command with the NEW command discussed earlier). The "filename" refers to whatever unique file name has been used to identify the particular file. Some other command might be used on your system. In any case, the command will cause the information in the permanent file to be "loaded" into the computer's main memory. Then you can use the file (run the program, edit the file, print the information, etc.).

The command(s) to retrieve a file that has previously been saved on my

system is _____

_____

Editing and Replacing Files

Frequently, files must be edited. New information must be added or inserted, corrections must be made, unwanted information must be deleted, and so on. Then the old version of the file must be replaced in permanent memory by the new edited version. Computers need to have the capability to perform these functions.

Editing Procedures

The procedures for editing files differ greatly from one computer to another. Thus your instructor will have to show you the exact procedures for your computer. Further, if you use a computer for a variety of purposes such as programming in BASIC, statistical analyses, and word processing, you may find it necessary to learn to use several editing systems.

On most computers, editing requires that you first load the permanent file into main memory. On some systems, you can immediately begin editing. On others, you must use another command to tell the computer that you wish to edit the file. In either case, there will be certain rules to follow in order to add, insert, delete, or move information in the file.

Some editing systems are cursor-oriented. Editing is performed at the current location of a cursor, a signal showing the user where the computer is located in the file. The user can move the cursor to any location (symbol, letter, digit, etc.) in the file and then type in new information or use special keys to delete unwanted information. Other systems are line-oriented. The user must have the computer find a particular line in the file and then give commands to change the line, insert new lines, delete the line, and so on.

Command(s) I must use to edit a file on my computer:

_____

_____

_____

My editing system is ☐ cursor-oriented, or ☐ line-oriented (Check one.)

Is there a command to stop editing? ☐ yes ☐ no

If yes, the command is _____

Other information on my editing system regarding adding, inserting, deleting, and modifying information:

_____

_____

_____

Replacing a File
After a file has been changed, the user will usually wish to replace the current permanent file with the new version. That is, the edited file

will take the place of the original one in permanent memory. A typical command to do this is

REPLACE,filename

although, depending on the computer, the command may simply be the same as that used originally to save a file, or a command within the editing system may be used.

The command(s) used to replace a file on my computer:

_____

_____

Other information on file replacement:

_____

_____

_____

Removing a File

When a permanent file is no longer needed, it can be erased from the tape or disk. This frees up the space on the auxiliary memory device to allow storage of new information. Your computer has a command to allow you to do this. Typical commands include PURGE, REMOVE, DELETE, and UNSAVE. For example, the command

PURGE,filename

would tell the computer to remove from permanent memory (purge) the contents of the file identified by the file name.

The command to remove a file on my computer is _____

_____

Back-Up Systems

It takes a great deal of time and effort to input information to the computer. In economic terms, this is a cost. Any user begins to feel uneasy about the consequences of the information somehow getting lost or destroyed. This can happen to a user on any system from a home microcomputer to the largest and most sophisticated mainframe computer. To illustrate the problem, I can relate a story about a graduate student who collected data for her thesis from over 200 condominium residents. Each resident provided information on over 100 variables concerning his or her dwelling and his or her satisfaction with the dwelling. The data from each resident required that the student type in four lines of numbers on the computer. Thus the entire data file consisted of over 800 lines of numbers. One day the student came in to conduct statistical analyses and discovered that the file consisted of thousands of lines of meaningless characters generated through some sort of computer malfunction. Many computer users can provide their own horror stories of information lost to fire, rain, sun, or whatever.

To the great relief of my student, the computer center had a back-up system. Each day, all users' files are copied on tape and stored in a safe location. The original data file was located on one of these tapes, and the information was loaded onto the disk for her use. Every computer system must have some procedure for back-up to protect the user from accidental loss of files. Even home computer users make copies of their small floppy disks.

Summary

Information on a computer is often organized in the form of files. Computer users construct either data files or program files and then instruct computers to process the information in the files. All computers must have some method for allowing the user to construct files.

Each computer system uses its own instructions for file construction and editing. In all cases, however, the computer will have a method to allow the user to

- Create a new file
- Save a file in permanent memory

- Retrieve a file from permanent memory and load the file into the computer's main memory
- Edit the file to make corrections, add or delete information, etc.
- Replace the old version of the file with the edited version in the computer's permanent memory
- Remove or purge a file from permanent memory
- Allow the user to have access to a back-up copy of the file in case the original file is destroyed

Further Reading

The manuals on the operating system of the computer you are using contain a great deal of information on file management procedures with your computer (including more sophisticated applications than those described in this chapter). Your computer center may also have a user's manual that summarizes the data input and storage procedures on your system.

3

Statistical Analysis
with
Computers

Social and behavioral scientists have found that one of the great benefits of computer technology is the ability to quickly and easily perform statistical analyses that are time-consuming and difficult to carry out by hand or even on a calculator. Computers are perfect "number crunchers" and easily adapt to the task of calculating complex statistical formulas. Chapters 4, 5, and 6 will describe several of the commonly available statistical packages that are designed to perform statistical analyses of scientific data and will give the fundamental procedures for using each of them. This chapter will serve as a general introduction to using these statistical tools.

Statistical Packages

Statistical software packages are commercially available computer programs written to perform statistical analyses on data input by the user. The programs are written in a computer language such as BASIC or FORTRAN (see Appendix). Although a number of packages are available, you will have to find out which ones are available on your campus.

The most widely used packages are SPSS (Statistical Package for the Social Sciences), Minitab, BMDP (BioMeDical Computer Programs), and SAS (Statistical Analysis System). Your campus computer system probably has one or more of these packages. Your computer may also have available special programs designed to perform very specific computational tasks. These may be for statistical analyses or business applications such as loan amortization and business forecasting. Statistical packages are also commercially available for microcomputers.

Computation
Versus Statistical Theory

Computers are extremely useful for performing statistical tests. They are fast and they are accurate. Yet they are only a tool that saves researchers from the burden of hand calculation. In other words, computers are useful for computation, but they cannot tell a researcher why a particular test is appropriate or how to interpret statistical results. Computers also have no way of knowing whether a particular research design is free of methodological flaws. The computer user needs to have the necessary knowledge of statistical theory and research methods to use the computer as a tool for computation of statistical tests. Neither the computer program nor this book will be able to teach you statistical theory and research methods.

"Garbage in, garbage out" is a popular saying among computer users. Unless the computer input is meaningful, the output will be meaningless. The saying is particularly appropriate when applied to statistical analysis of research results. The lesson here is that a knowledge of statistics and research design is extremely important.

Using Statistical Packages

This section will describe some of the general features of the various statistical packages. The first thing to remember is that the package is a computer program that must be accessed in order for you to use it. Your computer will have a specific command that will access the package. Once this is done, you will operate according to the rules and procedures of the particular package. For each package, you must learn the rules for instructing the package to perform a statistical analysis. When the statistical analyses have been completed, each package will have a command to exit from the program. When this is done, you are able to use the computer for other purposes (run other programs, construct new files, etc.).

Chapters 4, 5, and 6 provide simple data analyses as examples, along with sample computer output. You are encouraged to use the data provided to run programs on your computer because hands-on experience is the best way to learn about computers. Please keep in mind, though, that the example data provided are all fictitious—that is, they were all constructed only for the purpose of illustrating the computer packages.

On-Line Versus Punched Card Systems

Your work with the computer will be either in an on-line "interactive" mode or with punched cards ("batch" mode). During on-line interaction with the computer, the user's command to the computer is processed immediately, and a new command may be given right away. When using a punched card system, all data and instructions to the computer are typed on computer cards. These are placed in a card reader that is attached to the computer. Eventually, the output is printed and the user picks up the output (usually somewhere in the computer center). Thus feedback is not as immediate. If a different analysis is desired, new cards must be punched and inserted into the computer card deck, and the deck must again be fed through the card reader.

The procedures you will follow when using a statistical package will differ somewhat depending upon whether you are using an interactive or punched card system. This book will not attempt to give you every difference in detail. It is assumed that you will be able to acquire knowledge of the exact procedures for running a statistical package on your particular computer from your instructor.

Note: The remainder of this chapter is intended for students with a limited background in statistics. It may be omitted by anyone who has a fundamental knowledge of statistical procedures.

Data Input

Studies in the social and behavioral sciences are designed to answer questions such as

- Do people like others with similar attitudes more than others with dissimilar attitudes?
- Does a particular teaching method produce better learning than some other method?
- Is crowding associated with crime?   •
- Does viewing television violence lead to increased aggressiveness?

These questions all ask about the nature of certain variables. A variable is a general class or category of objects, events, situations, or personal characteristics. Within this general class, specific instances are found to vary. Thus some of the variables in the questions above are attitude similarity, liking, type of teaching method, and crowding.

Studies conducted to answer such questions involve collecting data from subjects on the relevant variables. For example, in a study on attitude similarity and liking, a researcher might have 50 subjects meet someone with dissimilar attitudes and another 50 subjects meet someone with similar attitudes. The researcher would then collect data on the liking variable by asking subjects to indicate how much they liked the other person.

The results of the study must be analyzed, and this is the point at which computers become useful tools. To be able to perform the analysis, data must be input. All of the statistical packages require the data to be input in the form of columns such that each column corresponds to a particular variable or level of a variable. For example, the attitude similarity data could be analyzed using Minitab with two columns of data. One column would contain liking scores for all the subjects in the dissimilar attitude group, and the other column would contain liking scores for all subjects in the similar attitude group.

Data Analysis Capabilities

Once data input has been completed, the researcher uses commands of the statistical package to accomplish the desired data analyses. A brief introduction to some of the data analysis capabilities of computers is provided here.

Descriptive Statistics

Descriptive statistics provide a description of the data. All packages will provide these. The statistics include

Mean: the arithmetic average
Median: the middle score in a group
Mode: the most frequent score
Variance: the variability of scores about a mean
Standard deviation: the average deviation of scores about a mean
    (square root of the variance)

Statistical packages also have the ability to provide frequency distributions and graphic displays such as histograms (bar graphs) and plots showing the relationship between variables.

Correlation

All packages calculate correlation coefficients. They will also perform regression analyses and provide regression equations. A correlation coefficient is an index of the strength of the relationship between two variables. The correlation may range from −1.00 to +1.00. The + and − indicate whether the relationship is positive or negative. The nearer the correlation is to ±1.00 the stronger the relationship. For example, studies of the intelligence test scores of identical twins reared together show a strong correlation (about +.90) between the two scores. In contrast, the scores of nontwin siblings reared together are correlated less strongly (about +.60). The scores of unrelated children reared together correlate about +.35, and scores of unrelated individuals who did not live together are not correlated at all (0.00).

Cross-Tabulation

All packages construct cross-tabulated frequency distributions (contingency tables). These show frequencies as a function of two variables. For example, a study might look at the frequency of people who did or did not vote as a function of their political affiliation.

Statistical Significance Tests

All packages contain a variety of tests to inform the researcher of the statistical significance of the results of a study. To illustrate, suppose the researcher in the attitude similarity study finds that the mean liking score is higher in the similar than in the dissimilar condition. Because these means are based on only one sample of individuals, can the researcher infer that this difference is a true one that would hold up if an entire population were studied or if the study were conducted many times, each time with a new sample of subjects? A statistical significance test gives the probability that the result obtained with this sample is simply a fluke. A result is said to be significant when there is a low probability (usually .05 or less) that the result is due to some random occurrence in the particular sample studied. Most packages will perform tests such as a t-test, analysis of variance, Chi-square, and various nonparametric tests.

Data Modification and Selection

Statistical packages allow a variety of data modification or selection. For example, some of the possible manipulations include

- Transforming scores on a variable by squaring them, taking the square root, or taking their absolute value
- Computing new variables by adding, multiplying, subtracting, or dividing scores on two or more variables
- Selecting certain subjects for analysis on the basis of some selection criterion
- Recoding the values of a variable

Conclusion

No single book can fully explain and describe all of the available statistical packages. Each package has complete manuals that give exact details on all of the capabilities of the package. References to the various manuals for the most commonly used packages will be provided when those packages are described in subsequent chapters. Also, although this book can provide the fundamental information that will allow you to use a package, all of these packages can do much more in terms of sophisticated data analysis than can be described here. Finally, there is no substitute for hands-on experience; the best way to learn about a statistical package is to use it, experiment with it, make mistakes, and try to figure out why you made those mistakes. Remember that using computers is a problem-solving activity—there is always a solution, and computer users must always approach computer usage as a challenge that can be very rewarding when success is achieved.

Summary

A major use of computers in the social sciences is statistical analysis. Computers can perform mathematical calculations very quickly and reduce the errors that are introduced by manual calculations. Statistical packages for computers are commercially available programs that perform the statistical analyses.

The most widely used statistical packages are SPSS, Minitab, BMDP, and SAS. Subsequent chapters will describe these. Your campus probably has other statistical programs, and programs may be purchased for microcomputers.

Statistical packages are only tools to simplify a tedious chore. They are not a substitute for a thorough knowledge of research methods and statistical theory.

Each statistical package has its own rules and language for operation. These must be learned in order to use the package. The exact procedures will also differ depending upon whether the system is being used in an on-line interactive mode or a punched card batch mode.

All statistical packages require that data files be constructed. Once data are input to the computer, the user must instruct the computer to perform the desired statistical analyses. Commonly available statistical calculations include

- Descriptive statistics
- Correlation
- Cross-tabulation
- Statistical significance tests such as a t-test, analysis of variance, Chi-square, and so on
- Data modification and selection

Further Reading

Cozby, P. C. (1981). *Methods in behavioral research.* Palo Alto, CA: Mayfield Publishing Co.

Kidder, L. H. (1981). *Seltiz, Wrightsman & Cook's research methods in social relations.* New York: Holt, Rinehart & Winston.

4

Statistical Packages I:
SPSS

SPSS is the *Statistical Package for the Social Sciences.* If you use SPSS, you will probably have one of the following available to you: (1) SPSS on-line system, (2) SPSS punched card system, (3) SCSS conversational system.[1] SCSS is a relatively new package that uses a conversational mode in which the computer asks the user questions. In the process of answering, the data analysis is accomplished. The SPSS corporation has also recently introduced an enhanced version called SPSS[x] and has made available a graphics option. This chapter will describe the SPSS system as used with punched cards or on-line with a computer terminal. The material in this chapter will probably have to be supplemented by material from your instructor because the exact version of SPSS on your computer may have some special characteristics that you will need to know about.

There are two steps in the use of SPSS. The first is constructing a data file. The second is writing an SPSS program to perform the statistical analysis of the data. When the SPSS program is executed in conjunction with the data file, the analyses will be output.

The Data File

The data file consists of a series of lines of data (or punched cards on a card system). Research investigations will require that some number of subjects provide data on some number of variables. For example, a researcher might collect data from 200 subjects on sex, income, education, religious preference, and marital satisfaction (a total of five variables). In an SPSS data file, the data for each subject is formatted in columns across the line. Thus the first line (or card) would contain data provided by the first subject on the variables being studied. The second line would contain data from the next subject, and so on. In our example, the data file would consist of 200 lines; each line would contain a series of numbers representing a subject's data on the variables being studied. This is illustrated in Figure 4–1. The data file itself consists only of data needed for the analysis; it does not include the names of the actual variables or any other information.

Numeric Representation of Data

Computer analysis requires that all data be coded numerically.[2] Even though a subject's responses might have been something like yes or no, these must be coded in the computer as numbers—for example, a 1 if the subject said yes and a 2 if the subject said no. The same principle

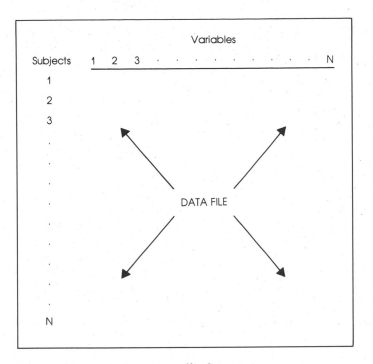

Figure 4–1   SPSS Data File Structure

applies to all variables in a study. In our example, sex would have to be coded numerically (e.g., 1 = male, 2 = female) as would all of the other variables. Religious preference might be coded as 1 = Protestant, 2 = Catholic, 3 = Jewish, etc.

Note that for variables such as sex or religious preference, a number had to be assigned to each possible response even though the numbers themselves don't have any true numeric qualities (i.e., in the code for sex, a 2 does not imply anything greater than a 1). For other variables, such as income, the numbers have true numeric qualities, and it is obvious that numeric coding of the data would be used. However, the coding system would be determined by the response format. If the researcher asked for exact yearly income, someone who responded "$15,826" would be coded as 15826 whereas someone who said "$15,829" would be coded as 15829. Note that commas and dollar signs are *never* used in coding data.

Two further points about the income variable example may be made. First, you might wonder what to do if the researcher had asked for income information in the following way:

Check one of the following:

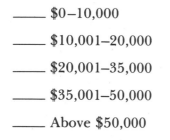

_____ $0–10,000

_____ $10,001–20,000

_____ $20,001–35,000

_____ $35,001–50,000

_____ Above $50,000

In this case, the logical coding scheme would be to code the first possible response ($0–10,000) as 1, the second as 2, and so on. Second, you might wonder whether a researcher might want to code the data so that all persons within a certain income range receive the same numeric code (e.g., the persons who said $15,826 and $15,829 would both be coded with the same number because both incomes are within the same range). It is possible to do this when the researcher is coding the data and preparing the data file. However, it is possible for SPSS to do this automatically. As we will see, SPSS has the capability to recode numeric values.

Coding Missing Data

In many research studies, subjects may not provide data on one or more of the variables. For example, one subject might answer all the questions except religious preference, and another might not indicate income. The reasons for missing data vary across studies and individuals. It is of course possible to simply exclude from the data file any subject who has missing data. However, most researchers prefer to assign a special code for missing data so that the rest of the subject's data may be analyzed. SPSS allows the researcher to instruct the computer to consider one or more values on a variable as "missing" and not include missing values in the statistical computations.

The numeric code for missing data must be different from the numbers used for real data. The easiest method is to leave a blank space so that a blank will be considered "missing" whereas any number will be considered "real" data. Another approach is to use a number that could not possibly be a real value. For example, the values for the variable of amount of education might range from 0 (no education) to 16 (16 years of school). Here, a researcher might use a value such as 99 as the missing data code.

It is also possible to use more than one missing value. Sometimes it is useful to do this in order to distinguish between or among the reasons why there is missing data. For example, different codes could be used depending on whether the subject refused to answer or didn't know the answer, or the interviewer forgot to ask the question.

Coding Decimal Point Data

Sometimes numeric data are expressed in fractions of whole numbers. A student's grade point average (GPA) usually can range from 0.00 to 4.00, with the latter number representing a GPA consisting of all "A" grades. When collecting data, the researcher would obtain actual GPA values such as 2.13, 3.00, and 3.67.

How should such data be coded in the data file? Although it is possible to type in the numbers and the decimal points, it is much easier and more efficient to code these numbers *without* the decimal point. Thus the GPA values above would be coded in the data file as 213, 300, and 367. The SPSS program that is written to do the analysis will be instructed to automatically place the decimal in the proper place when analyzing the GPA variable.

Formatting Data in Columns

The material presented thus far has concerned the *coding* of data for the data file. At this point, the data may be entered in the data file. The data for each subject should be placed in columns on a line in the data file. The most important thing to remember is that each variable should be coded in the exact same column for every subject. An example should help make these points clear.

Suppose that you are going to input the data for the first five subjects in a study in which the variables were sex, income, education, religious preference, and marital satisfaction. You have determined the numeric code for each variable and decided that blanks will be used for missing data. Now suppose that for these five subjects, you obtained the following data:

Subject 1:   1, 15287, 12, 2, 7
Subject 2:   2, 15289, 13, 1, 4
Subject 3:   1, 25100, 16, missing, 3
Subject 4:   1, missing, 8, 1, 6
Subject 5:   2, 9500, 14, 3, 5

In the data file, the values of the variables must be placed in exact columns on the line, and the variables are entered in the same order for every subject. What would the data file look like? One possible format for the data file is shown here; a subject number is used at the beginning of each line in the file.

```
001  1  15287  12  2  7
002  2  15289  13  1  4
003  1  25100  16     3
004  1          8  1  6
005  2   9500  14  3  5
```

There are several things to note about this data file. First, a subject number was used although this is not mandatory or absolutely neces- sary. However, it is frequently convenient to have subject numbers when editing a file and in certain other situations as well.

Also note that a certain number of columns are used for each variable. The number of columns assigned to a variable must be sufficient to accommodate all possible values of the variable. If values can range from 1 to 9, then only one column is necessary. Two columns are needed for values up to 99, three for values up to 999, and so on. If five columns are assigned to income, the data file can contain values up to $99,999. If it is possible for any subject to have higher income, then six columns would have to be used. If negative numbers are used (e.g., −1 or −2) a column must be reserved for the minus sign. These details must be planned out prior to preparing the data file.

It is also important to remember that the columns assigned to a variable follow conventional numeric notation in which the column farthest to the right is the "ones" place, the next column to the left is the "tens" place, and so on. Note the placement of the 8 for subject 004. The value must go in that column to be read as "8" by the computer. If it had been placed in the other column (directly below the 1 in 16), the computer would read the value as "80" instead of "8." The same principle may be observed in the value for income for subject 005.

Finally, you should note that in the data file example, a blank column was left between the columns used for the actual variables. This was done solely for the purpose of visually separating the vari- ables in the data file. The only advantage of doing this is to make it somewhat easier to check the data file for errors. (It is extremely important that researchers make sure that the numbers entered into the data file are correct.) However, the blank columns are not necessary, and the following data file would be perfectly acceptable to the computer (and would be functionally equivalent to the data file shown above).

```
0011152871227
0022152891314
00312510016 3
0041      816
0052 95001435
```

One important point: Unless you have a good reason for doing so, never leave more than one blank space between the data columns in your data file.

If you are using a computer terminal, the procedures for input and saving of your data file will be the same as those described in Chapter 2. If you are using a punched card system, your data file will consist of a deck of cards with the data punched on the cards.

A data file that expands upon the small example given on page 44 is shown in Box 4–1. This data file will be used for most of the output that is displayed later in this chapter.

The procedures for formatting data described so far have emphasized placing data in a specific set of columns. This is called fixed-column data formatting. The fixed-column method is generally the more common method used by researchers. However, another method called free-field formatting is available. You should check with your instructor about whether you can use the free-field method. If free field is used, blanks cannot be used as missing values; some specific numeric value must be designated as the missing value.

Multiple Lines for Each Subject

Sometimes all the data for each subject cannot be entered on one line of the computer terminal or on one punched card. In much research, a great deal of data is collected from each subject. Sometimes a questionnaire or interview may contain many questions, or a psychological test may have several hundred items. Researchers may also wish to go to a new line whenever a new portion of a questionnaire is input (e.g., one line for personal background, another for political views, another for occupational information, etc.). SPSS easily accommodates such situations. If it is necessary to use more than one line or card to input all the data for a subject, the researcher simply continues the data input on the next line or card in the data file.

Box 4–1   A Data File for SPSS Examples

A data file that will be used to generate the output for most of the SPSS statistical procedures described in this chapter is shown here. You may find it useful to enter this file in your computer and use it with SPSS programs that you write. The data for each subject occupy 18 columns. There are six variables: subject number, sex, income, education, religious preference, and marital satisfaction. The format is the same as the example data file described in the text.

001	1	15287	12	2	7
002	2	15289	13	1	4
003	1	25100	16		3
004	1		8	1	6
005	2	9500	14	3	5
006	1	28200	15	1	5
007	1	41100	17	1	7
008	1	8666	9	1	4
009	2	12617	12	1	1
010	1	15190	12	2	2
011	2	29900	16	2	3
012	2	31000	16	1	3
013	2	26500	17	3	7
014	2	9000	10	2	4
015	1	41555	17	1	6
016	1	55000	17	1	6
017	1	6400	10	4	2
018	2	29500		4	7
019	2	28000	16	1	7
020	1	34000	16	1	
021	1	12000	14	1	4
022	2	10200	12	2	3
023	1	31500	16	2	5
024	1	36000	16	4	5
025	1	15000	12	4	5
026	1	11220	10	1	4
027	1	16000	13	1	2
028	2	18500	13	2	1
029	2	10000	12	2	7
030	2	20000	12	1	5
031	2	24500	14	1	5
032	2	31000	16	3	6
033	2	35500	16	1	6
034	1	21000	16	3	2
035	1	9500	12	2	3
036	1	25650	15	1	5
037	1	18200	12	1	6
038	2	29000	17	2	6
039	2	21200	16	1	6
040	2	30500	16	3	7
041	2	36000	17	2	7
042	2	30000	17	1	7
043	2	32200	15	1	7
044	2	21200	14	1	6
045	2	18760	12	2	5
046	2	36600	17	2	5
047	1	29000	15	1	6
048	1	21500	16	1	6
049	1	22900	12	2	6
050	1	26000	16	2	5

When two or more lines are used, it is important that the data file contain the same number of lines for each subject. If, for example, it is necessary to use two lines for most subjects, then every subject must have two lines *even if no data is input for a subject on the second line.*

Coding Independent Variables

Research designs in which subjects are assigned to two or more groups on an independent variable must have a column in the data file that contains group identification numbers. For example, suppose one group of subjects is exposed to high-level noise and a second group receives low-level noise (noise level is the independent variable). All subjects are then measured on a reaction-time task (the time it takes to make a reaction such as pushing a button is the dependent variable). This design would require that one column in the data file contain a number to identify whether the subject was in the low- or high-noise group (usually either 1 or 2). A second data column would contain the score on the reaction-time dependent variable. Factorial designs in which there are two or more independent variables require that a separate column be used for group identification on each of the independent variables.

The SPSS Program

After a data file has been prepared, an SPSS program must be written to enable the computer to perform the analyses desired by the user. SPSS programs consist of a series of *commands* that perform a general function, such as performing a statistical test or recoding a variable. Each command is followed by the specific instructions that pertain to the user's data file—for example, to perform the statistical test on a particular variable.

The method you will use to write an SPSS program depends upon whether you are using an on-line or a punched card system. We can illustrate this difference with the RUN NAME command. RUN NAME is an optional command that functions to print a name at the top of each page of output. It is a convenience that helps the user remember the nature of the data in the printouts. The following sections on using the on-line and card systems should both be read.

On-Line SPSS

To input a program using an on-line system, you must first tell the computer to access SPSS. When this is done, SPSS will ask some questions about your data file name and then give you a prompt that signifies that you are ready to input the first line of your program. Your instructor will show you the exact procedures for accessing SPSS. Note these in the spaces provided. (*Note:* On some computers, the procedure to input and run a program will differ from that presented here. Follow your instructor's directions.)

The command to access SPSS on my system is _____
SPSS then asks the following questions, and I then answer with the following responses:

Question_____

Response _____

Question _____

Response _____

What tells me that I am ready to input my SPSS program?

_____

_____

The SPSS program consists of a series of line numbers. The usual convention is to begin with line 10 and increment lines in steps of five (10, 15, 20, and so on). This allows insertion of lines later on if necessary.

When the RUN NAME command is used, it is the first line of the program. Suppose you want the name ATTITUDE STUDY to be printed at the top of each page (up to 64 characters including blank spaces may be specified).[4] When the prompt appears, you would type the command and the name you want to use. The command RUN NAME would be typed on line 10. The specific name would be typed on a "subline." Sublines take the form of 10.005, 10.010, 10.015, and so on. Thus 10.005 is the first subline of primary line number 10. This sounds complicated but in practice it is simple. You would simply type two lines (each followed by hitting the return key):

```
10 RUN NAME
10.005 ATTITUDE STUDY
```

In actuality, typing the program is even simpler. One line could have been typed:

```
10 RUN NAME;ATTITUDE STUDY
```

The semicolon following the command tells SPSS to consider subsequent information as the first subline (10.005).

There are several other features of on-line SPSS you will need to know about:

1. To list your program on your terminal, simply type LIST or L (no line number is necessary because this is not part of the SPSS program). To list only a series of lines, type L followed by the line numbers. For example, typing L,10,50 would list lines 10 through 50. A single line number can be listed also. For example, L,10 would list line 10 only.

2. SPSS executes the program in order from the lowest to highest line number. SPSS will automatically insert a line at its proper point no matter when you type it. Thus, if you type in line number 5 after typing line 10, SPSS will place line 5 *before* line 10.

3. If you use a line number that is currently in the program, the new input will automatically replace the old line. This is useful for correcting errors in a line. Your instructor may also have information on editing methods that allow you to correct errors made on a line in your SPSS program.

4. You can easily delete lines from the program. Typing D,10 would delete line 10. Typing D,10,60 would delete lines 10 through 60.

5. If the specifications require that you use more than one line, you simply add sublines with numbers such as 10.010, 10.015, 10.020, 10.025, and so on.

Punched Card SPSS

Punched cards allow the user to type in 80 columns of information. In SPSS, the command is typed beginning with the first column. The specification for the command is typed beginning in column 16 on the card. The SPSS program consists of a series of cards containing command instructions and the exact specifications for the data.

If the RUN NAME command is used, it will normally be the first card in the program. The command would be typed on the card. If we want ATTITUDE STUDY printed on each page of output, that would be typed beginning in column 16. The card would appear as in Figure 4–2. Changing an SPSS punched card program requires that cards be manually added or removed before the card deck is run through the card reader again.

Your instructor will provide the exact instructions you will need to run your program, including any differences from the on-line system,

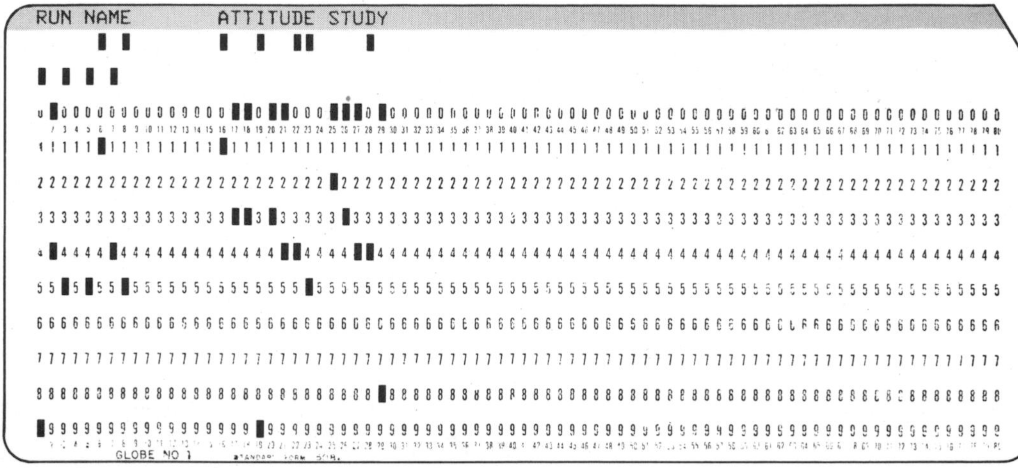

Figure 4–2   Punched Card for RUN NAME Attitude Study

which we will focus on in the remainder of this chapter. Punched card users especially need to know about the INPUT MEDIUM and READ INPUT DATA cards.

Data Definition Commands

Suppose in writing your SPSS program you began on line 10 with a RUN NAME optional command. The next lines would be data definition commands that in essence define your data file for the computer. These include VARIABLE LIST, INPUT FORMAT, N OF CASES, VAR LABELS, VALUE LABELS, and MISSING VALUES.

VARIABLE LIST (Required)
The VARIABLE LIST command is used to name the variables in the data file. Each variable name may be up to eight characters in length and must begin with an alphabetic character. An example of a variable list for the data file we described earlier would be

    VARIABLE LIST;SUBNUM,SEX,INCOME,EDUC,RELIG,MARSAT

Note that the variables are listed in the same order in which they appear in the data file. They are abbreviated names because they cannot exceed eight characters. The names assigned to the variables must be used throughout the SPSS program. Also note that each variable name is separated by a comma. An alternative method is to separate each of them with a single blank space. In the example above, the semicolon convention was used. If the variable list had been written on line 15, SPSS would understand the line as

15 VARIABLE LIST
15.005 SUBNUM,SEX,INCOME,EDUC,RELIG,MARSAT

If additional lines are needed to specify more variables, they would be written on lines 15.010, 15.015, etc.

A shortcut in naming the variables is to use the TO convention. A specification such as V1 TO V8 would name eight variables: V1, V2, V3, V4, V5, V6, V7, V8. The variable name may have up to five alphabetic characters and up to three numeric characters. Thus a variable list such as QUEST001 TO QUEST999 is possible. (*Note:* SPSS statisical commands may have limits on the number of variables that can be analyzed.)

INPUT FORMAT (Required)

The INPUT FORMAT command is used to tell SPSS where each piece of data is located in the data file. It is used by the computer in conjunction with the VARIABLE LIST. Whereas VARIABLE LIST names the variables, INPUT FORMAT specifies the columns in which the variables are located in the file. The INPUT FORMAT for the data file example we have been using would be

INPUT FORMAT;(F3.0,1X,F1.0,1X,F5.0,1X,F2.0,1X,F1.0,1X,F1.0)

Some SPSS systems require the use of the word FIXED to indicate fixed-column format:

INPUT FORMAT;FIXED(F3.0,1X,F1.0,1X,F5.0,1X,F2.0,1X,F1.0,1X,F1.0)

Check with your instructor regarding whether this is needed.

Let's examine the purpose of each element in the input format statement above.

F3.0	The first variable is located in the first three columns of the data file. This corresponds to SUBNUM in the variable list. The ".0" indicates that there are no decimal spaces.
1X	The "X" is used to indicate blank columns. 1X specifies a single blank column that will be ignored by the computer.
F1.0	The next column contains the second variable, SEX.
1X	Another blank column.
F5.0	The next variable is located in the next five columns. This would correspond to INCOME on the variable list.
1X	Another blank column.

F2.0   The next two columns contain the data for EDUC.

1X   A blank column follows.

F1.0   One column for RELIG.

1X   A blank column.

F1.0   This column contains the MARSAT score.

Note that six variables were specified in the variable list and the input format. It is a good idea always to check that these correspond. Also, the input format accounted for 18 columns. It is good practice to examine your data file to make sure that your input format specifies that the correct number of columns were used. Finally, there must be an equal number of left and right parentheses. If these procedures are not followed correctly, SPSS will give an error message when the program is executed. Simply double-check the correspondence between the data file, the VARIABLE LIST, and the INPUT FORMAT, and make any necessary corrections.

It is possible to write your input format in a simplified manner if there are repeating patterns in the use of columns. For example, 10F1.0 would specify that there are 10 variables each having one column and with no spaces in between. As another example, the following specifications might be used:

(F3.0,10(1X,F1.0))
(F3.0,5(F1.0,F2.0))

The first format has the expression 10(1X,F1.0), specifying that there is a repeating pattern of one blank space and a one-column variable; the pattern repeats ten times. The second expression 5(F1.0,F2.0) specifies that a pattern of a one-column variable followed by a two-column variable repeats itself five times. In both these examples, the number of left and right parentheses are equal.

All of the examples we have been discussing specify variables without decimal places (all are F1.0, F2.0, etc.). To indicate decimal points, expressions such as F2.1 or F3.2 are used. F2.1 means that the variable occupies two columns with one decimal point place—in the data file a number such as 34 would be read as 3.4. F3.2 indicates that three columns are used for the variable and that there are two decimal places. Thus a GPA score coded as 267 would be read as 2.67 by SPSS. (*Note:* If your data file does include variables with decimal points, you should read about the PRINT FORMATS keyword in the SPSS manual.)

If data for a subject require that more than one line per subject is used in the data file, a slash must be used in the input format to tell

SPSS when to skip to the next line. For example, (2OF2.0/30F1.0) would inform the computer that there are 20 variables on the first line, each occupying two columns; the slash directs the computer to the next line to find 30 more variables. The variable list would indicate that there are 50 variables for each subject.

The free-field format was mentioned earlier. If you use this on your system, the INPUT FORMAT statement becomes extremely simple:

INPUT FORMAT;FREEFIELD

N OF CASES (Optional with On-Line SPSS)
The N OF CASES command specifies the number of subjects in the data file. If there are 50 subjects in a study, the instruction would simply be

N OF CASES;50

This would be the next line in the program. A shortcut in specifying the number of cases is to use UNKNOWN. The following line will cause SPSS to count to the bottom of the data file to provide the actual number of cases:

N OF CASES;UNKNOWN

If you know the exact number of cases in your data file, it is better to specify this on the N OF CASES line rather than use the UNKNOWN convention. Finally, although N OF CASES is considered optional, it is usually desirable to include it in the program.

DATA LIST (Alternative)
The DATA LIST command may be used as an alternative to VARIABLE LIST and INPUT FORMAT in the SPSS program. The data list contains both the variable names and the format of the variables in the data file. You may wish to consult the SPSS manual and experiment with the DATA LIST command (also see the material on SPSS[x] below).

VAR LABELS (Optional)
The VAR LABELS command allows the user to assign extended labels to the variables named in the variable list. Variable names are limited to eight contiguous characters. By using VAR LABELS, a variable label of up to 40 characters can be assigned to any of the variables in the study. Here is an example of the VAR LABELS command:

VAR LABELS;SEX,SEX OF SUBJECT/RELIG,RELIGIOUS PREFERENCE/

This line would assign extended variable labels to the SEX and RELIG variables. The variable procedure involves typing the variable name and the variable label separated by a comma. If there is more than one variable label, each is separated by a slash.

VALUE LABELS (Optional)

The VALUE LABELS command is used to assign labels to the values of one or more variables. Each value label may be up to 20 characters in length. Here is an example of the value labels keyword:

    VALUE LABELS;SEX(1)MALE(2)FEMALE/MARSAT(1)LOW(7)HIGH/

This would assign a label to both values (1 and 2) of the sex variable and to two of the values of the marital satisfaction variable. As many variables as you like may be given value labels. It is also possible to use the same label for many variables by first specifying the names of the variables and then the value labels to be used. If all variables in the variable list are to have the same value labels, then the term ALL can be used. For example:

    VALUE LABELS;ALL(1)STRONGLY AGREE(5)STRONGLY DISAGREE

All of these examples have used only one line for the value labels. For most uses, you will have to continue the value labels on several sub-lines.

MISSING VALUES (Optional)

The MISSING VALUES command is used to inform SPSS that certain values of one or more variables are to be considered missing. The following is an example that would apply to our data file:

    MISSING VALUES;SEX TO MARSAT(BLANK)/

The general format is to specify the variable names followed by the missing values within parentheses. More than one value may be considered missing for any variable.

Note that TO was used in the missing values statement. The TO convention is a useful way of telling SPSS to include all variables in the VARIABLE LIST that are in between the two named variables. In the above example, the missing values specifications would apply to the variables of SEX, MARSAT, and all others listed between these on the VARIABLE LIST statement.

Data Modification and Selection

There are several SPSS commands that allow data to be modified or certain data to be selected for analysis. All are optional and used only when needed.

RECODE (Optional)

The RECODE command allows values of a variable to be recoded to take on different values. There are two common reasons to do this. One reason is to assign a single value to a range of values—for example, income $0–10,000 might be recoded as 1, $10,001–20,000 recoded as 2, and so on. Another circumstance arises in studies in which questions are worded in both positive and negative directions. For example, suppose that on a 10-question attitude scale odd-numbered items are worded so that agreement indicates a favorable attitude (e.g., "There should be no criminal penalties for the use of marijuana"), whereas even-numbered items are worded so that agreement indicates an unfavorable attitude (e.g., "Persons who use marijuana are dangerous criminals"). In this case, the researcher would want to recode half the items so that agreement always means the same thing.

RECODE takes the following form:

    RECODE;Q1,Q3,Q5,Q7,Q9(1=5)(2=4)(4=2)(5=1)

For all the variables specified, the old value is recoded to become the new value. That is, for Q1, Q3, Q5, Q7, and Q9, a value of 1 becomes a 5, 2 becomes a 4, and so on. This could have been done by hand prior to coding the data, but that would be tedious and might introduce errors.

COMPUTE (Optional)

The COMPUTE command allows a new variable to be computed. The following is an example of a COMPUTE statement:

    COMPUTE;NEWVAR=Q1+Q2+Q3+Q4+Q5

In the example, a new variable called NEWVAR is computed by adding the values of existing variables—Q1, Q2, Q3, Q4, and Q5. The name of the new variable must be different from any existing variable and must follow the same eight-character rule described earlier. Only one new variable can be computed on a line, but any number of new variables may be computed in an SPSS program.

All of the normal arithmetic expressions can be used to create new variables. Also, special functions may be used. The symbols used for some of these and accompanying examples are shown here.

Symbol	Use	Example
+	Addition	NV=Q1+Q2
−	Subtraction	NV=Q1–Q2
*	Multiplication	NV=Q1*Q2
/	Division	NV–Q1/Q2
**	Exponentiation	NV=Q1**2
SQRT	Square root	NV=SQRT(Q1)
ABS	Absolute value	NV=ABS(Q1)
COS	Cosine	NV=COS(Q1)

One note of caution if you use the COMPUTE command in your program: If the variables involved in the computation have missing values, you will need to use a command called ASSIGN MISSING. See the SPSS manual for details on this command.[5]

SELECT IF (Optional)

SELECT IF enables the user to select a certain subgroup of subjects to be included in subsequent statistical analyses. In the example data file we have been using, males could be selected with the following statement:

SELECT IF;(SEX EQ 1)

The expression within the parentheses directs the computer to sort through the data file and select only those subjects for whom the value of the sex variable equals 1. It is also possible to select cases that are greater than, less than, or not equal to a certain value. The following list shows the logical expressions used with SELECT IF.

SPSS Symbol	Expression	Example
EQ	Equal to	(SEX EQ 2)
GE˙	Greater than or equal to	(INCOME GE 20000)
LE	Less than or equal to	(EDUC LE 12)
GT	Greater than	(MARSAT GT 4)
LT	Less than	(INCOME LT 10000)
NE	Not equal to	(RELIG NE 1)

You can also select more than one variable using AND or OR. For example, the expression (SEX EQ 1 AND EDUC GT 12) would select males who had at least 13 years of education.

When a SELECT IF command is used, all statistics specified in the program will use only the selected subjects. It is used prior to the statistical commands with any other data transformation and selection procedure lines. Many times it is desirable to perform a *temporary* selection for the purpose of performing one statistical analysis. This may be done with a *SELECT IF command that is used in the section of the program that requests statistical analysis. The selection specified with *SELECT IF is in effect only for the statistic that immediately follows it. Thus *SELECT IF is called the "temporary select if" command.

SPSS Statistics

The remainder of this chapter will describe a few of the commands used to instruct SPSS to perform statistical analyses of data. There are more than 20 such commands. However, our main goal is to show you how to use a few of these so that you will have the background to consult the SPSS manual when you discover a need.

The general format for running statistical analyses is as follows:

100 STATISTIC COMMAND;VARIABLES SPECIFIED
110 OPTIONS;LIST
120 STATISTICS;LIST

As many statistical commands as necessary may be included. Most SPSS statistical commands have an associated set of options and statistics that can be selected by the user. Some of these will be presented here. For complete information on available options and statistics, consult your SPSS manual.

CONDESCRIPTIVE
CONDESCRIPTIVE performs descriptive statistics on continuous variables (that is, variables measured on a continuous number scale, such as income but not religious preference). The following are some examples:

CONDESCRIPTIVE;INCOME,EDUC,MARSAT
CONDESCRIPTIVE;V1 TO V20,V25,V50 TO V60
CONDESCRIPTIVE;ALL

Thus the list of variables may be an individual list, may use the TO convention, or every variable on the VARIABLE LIST may be included by means of the use of ALL.

The OPTIONS command normally is not used with CONDESCRIPTIVE. Option 3 writes out Z-scores, and option 4 prints an index.

Ten statistics are available. The following would produce the mean, standard deviation, and variance in the output:

    STATISTICS;1,5,6

All statistics will be output with the ALL convention:

    STATISTICS;ALL

An example of CONDESCRIPTIVE output is shown in Figure 4–3.

FREQUENCIES

FREQUENCIES produces a frequency distribution for one or more variables. The following are some examples:

    FREQUENCIES;GENERAL=SEX,EDUC,RELIG
    FREQUENCIES;GENERAL=ALL
    FREQUENCIES;GENERAL=V10 TO V20

Twelve options are available. Option 3 prints output in an 8 1/2″ × 11″ space, and option 8 prints a histogram. Eleven statistics may be specified, including

1. Mean
2. Standard error
3. Median
4. Mode
5. Standard deviation

The ALL convention can also be used for the statistics. The output of a FREQUENCIES run is shown in Figure 4–4.

CROSSTABS

CROSSTABS produces a cross-tabulated frequency table for two or more variables and calculates the appropriate Chi-square statistic. For example, a cross-tabulation of sex by education would yield a frequency table showing how many males and females are in each of the categories of education level. It can then be determined whether education

Figure 4–3 SPSS CONDESCRIPTIVE Example

The Program

```
10.      VARIABLE LIST
10.005 SUBNUM,SEX,INCOME,EDUC,RELIG,MARSAT
15.      INPUT FORMAT
15.005 (F3.0,1X,F1.0,1X,F5.0,1X,F2.0,1X,F1.0,1X,F1.0)
20.      N OF CASES
20.005 50
25.      VAR LABELS
25.005 SEX,SEX OF SUBJECT/RELIG,RELIGIOUS PREFERENCE
30.      VALUE LABELS
30.005 SEX(1)MALE(2)FEMALE/MARSAT(1)LOW(7)HIGH
35.      MISSING VALUES
35.005 SEX TO MARSAT(BLANK)
40.      CONDESCRIPTIVE
40.005 INCOME,EDUC,MARSAT
45.      STATISTICS
45.005 1,5,6
50.      STATISTICS
50.005 ALL
```

The Data File

See Box 4–1.

The Output

```
- - - CONDESCRIPTIVE - - -

00037000 CM NEEDED FOR CONDESCRIPTIVE



VARIABLE   INCOME

MEAN        23723.143    STD DEV   10435.665    VARIANCE   .109E+09

VALID CASES      49    MISSING CASES      1



VARIABLE   EDUC

MEAN          14.163    STD DEV       2.444    VARIANCE    5.973

VALID CASES      49    MISSING CASES      1



VARIABLE   MARSAT

MEAN           4.939    STD DEV       1.737    VARIANCE    3.017

VALID CASES      49    MISSING CASES      1
```

Figure 4—4   SPSS FREQUENCIES Example

The Program

```
10.      VARIABLE LIST
10.005  SUBNUM,SEX,INCOME,EDUC,RELIG,MARSAT
15.      INPUT FORMAT
15.005  (F3.0,1X,F1.0,1X,F5.0,1X,F2.0,1X,F1.0,1X,F1.0)
20.      N OF CASES
20.005  50
25.      VAR LABELS
25.005  SEX,SEX OF SUBJECT/RELIG,RELIGIOUS PREFERENCE
30.      VALUE LABELS
30.005  SEX(1)MALE(2)FEMALE/MARSAT(1)LOW(7)HIGH
35.      MISSING VALUES
35.005  SEX TO MARSAT(BLANK)
40.      FREQUENCIES
40.005  GENERAL=SEX,EDUC,RELIG
45.      OPTIONS
45.005  8
50.      STATISTICS
50.005  1,2,4,5
```

The Data File

See Box 4-1.

The Output

```
- - - FREQUENCIES - - -

FREQUENCIES - INITIAL CM ALLOWS FOR   1428 VALUES
              MAXIMUM CM ALLOWS FOR   8255 VALUES


SEX       SEX OF SUBJECT
```

CATEGORY LABEL	CODE	ABSOLUTE FREQ	RELATIVE FREQ (PCT)	ADJUSTED FREQ (PCT)	CUM FREQ (PCT)
MALE	1.	25	50.0	50.0	50.0
FEMALE	2.	25	50.0	50.0	100.0
	TOTAL	50	100.0	100.0	

```
SEX       SEX OF SUBJECT

      CODE
          I
      1.  ************************** (    25)
          I   MALE
          I
      2.  ************************** (    25)
          I   FEMALE
          I
          I........I........I........I........I........I
          0       10       20       30       40       50
          FREQUENCY

MEAN          1.500    STD ERR       .071    MODE        1.000
STD DEV        .505

VALID CASES     50     MISSING CASES     0
```

Figure 4–4   *Continued*

EDUC

CATEGORY LABEL	CODE	ABSOLUTE FREQ	RELATIVE FREQ (PCT)	ADJUSTED FREQ (PCT)	CUM FREQ (PCT)
	8.	1	2.0	2.0	2.0
	9.	1	2.0	2.0	4.1
	10.	3	6.0	6.1	10.2
	12.	11	22.0	22.4	32.7
	13.	3	6.0	6.1	38.8
	14.	4	8.0	8.2	46.9
	15.	4	8.0	8.2	55.1
	16.	14	28.0	28.6	83.7
	17.	8	16.0	16.3	100.0
	BLANK	1	2.0	MISSING	
	TOTAL	50	100.0	100.0	

EDUC

```
    CODE
         I
      8. **** (      1)
         I
         I
      9. **** (      1)
         I
         I
     10. ********* (      3)
         I
         I
     12. **************************** (     11)
         I
         I
     13. ********* (      3)
         I
         I
     14. ********** (      4)
         I
         I
     15. ********** (      4)
         I
         I
     16. ********************************** (     14)
         I
         I
     17. ******************* (      8)
         I
         I
  BLANK **** (      1)
(MISSING) I
         I
         I.........I.........I.........I.........I.........I
         0         4         8        12        16        20
         FREQUENCY
```

MEAN	14.163	STD ERR	.349	MODE	16.000
STD DEV	2.444				
VALID CASES	49	MISSING CASES	1		

(continued)

Figure 4–4 *Continued*

```
RELIG       RELIGIOUS PREFERENCE


                                                  RELATIVE   ADJUSTED    CUM
                                        ABSOLUTE     FREQ       FREQ     FREQ
        CATEGORY LABEL          CODE      FREQ      (PCT)      (PCT)    (PCT)

                                 1.        25        50.0       51.0     51.0

                                 2.        15        30.0       30.6     81.6

                                 3.         5        10.0       10.2     91.8

                                 4.         4         8.0        8.2    100.0

                               BLANK        1         2.0     MISSING
                                         ------      ------     ------
                               TOTAL       50       100.0      100.0
```

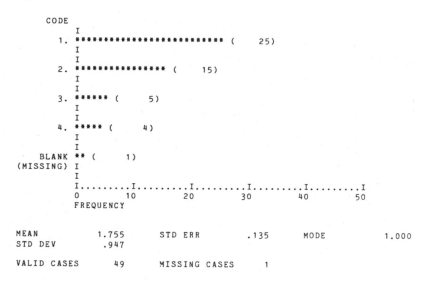

```
        RELIG       RELIGIOUS PREFERENCE

            CODE
                 I
            1.  ************************** (     25)
                 I
                 I
            2.  **************** (      15)
                 I
                 I
            3.  ****** (       5)
                 I
                 I
            4.  ***** (       4)
                 I
                 I
          BLANK  ** (       1)
        (MISSING) I
                 I
                 I.........I.........I.........I.........I.........I
                 0        10        20        30        40        50
                 FREQUENCY


        MEAN        1.755    STD ERR       .135      MODE       1.000
        STD DEV      .947

        VALID CASES    49    MISSING CASES    1
```

level varies as a function of subject sex. The CROSSTABS command takes this form:

CROSSTABS;TABLES=SEX BY EDUC

The variables specified for the crosstabs are separated with the word BY. Note also that TABLES= must be used.

There are 12 options, several of which serve to suppress printing of output that might not be desired. There are 11 statistics, including

1. Chi-square
3. Contingency coefficient
10. Eta

STATISTICS;ALL may be used as well. An example of the output of a CROSSTABS run with no options and all statistics is shown in Figure 4–5.

PEARSON CORR

PEARSON CORR calculates a Pearson correlation coefficient (r) between two or more variables. This statistic is used as an index of the strength of the relationship between two or more variables. Values of r may range from $+1.00$ to 0.00 to -1.00. A .00 correlation indicates no relationship (e.g., foot size and intelligence would not be expected to be related). A positive number indicates a positive relationship (e.g., the more hours spent studying, the higher the GPA) with values approaching $+1.00$ indicating a stronger degree of association between the variables. A negative number indicates a negative relationship (e.g., the more hours spent at social events, the lower the GPA); again, values approaching -1.00 indicate a stronger relationship.

The general format for this statistic is to use the command PEARSON CORR followed by a list of variables to be correlated. The ALL convention *cannot* be used. The following are examples of PEARSON CORR commands:

```
PEARSON CORR;EDUC,MARSAT
PEARSON CORR;INCOME,EDUC,MARSAT
PEARSON CORR;INCOME TO MARSAT
```

These commands would produce all possible correlations between the variables specified. The first would be only one correlation coefficient, but the second would produce three (INCOME with EDUC, INCOME with MARSAT, and EDUC with MARSAT) (*Note*: The TO convention may be used.)

The most useful option is number 6, which suppresses printing of "redundant" coefficients. In the first command above, only one correlation (EDUC with MARSAT) is desired. However, unless option 6 is used, the printout will contain EDUC with EDUC and MARSAT with MARSAT (a variable correlated with itself always produces a correlation of 1.00). There will also be MARSAT with EDUC; this is redundant because the correlation is exactly the same as EDUC with MARSAT. Usually statistics are not requested with PEARSON CORR. (Although statistic 1 prints out means and standard deviations, these are generally obtained with either CONDESCRIPTIVE or FREQUENCIES.)

Figure 4–5   SPSS CROSSTABS Example
The Program

```
10.      VARIABLE LIST
10.005 SUBNUM,SEX,INCOME,EDUC,RELIG,MARSAT
15.      INPUT FORMAT
15.005 (F3.0,1X,F1.0,1X,F5.0,1X,F2.0,1X,F1.0,1X,F1.0)
20.      N OF CASES
20.005 50
25.      VAR LABELS
25.005 SEX,SEX OF SUBJECT/RELIG,RELIGIOUS PREFERENCE
30.      VALUE LABELS
30.005 SEX(1)MALE(2)FEMALE/MARSAT(1)LOW(7)HIGH
35.      MISSING VALUES
35.005 SEX TO MARSAT(BLANK)
40.      CROSSTABS
40.005 TABLES=SEX BY EDUC
45.      STATISTICS
45.005 ALL
```

The Data File

See Box 4–1.

The Output

```
- - - CROSSTABS - - -

GIVEN 2 DIMENSIONS,   INITIAL CM ALLOWS FOR     502 CELLS
                      MAXIMUM CM ALLOWS FOR    4585 CELLS


       SEX       SEX OF SUBJECT
    BY EDUC


             EDUC
       COUNT
       ROW PCT                                               ROW
       COL PCT                                               TOTAL
       TOT PCT      8.        9.       10.      12.      13.
    SEX
          1.        1         1         2        6         1        25
    MALE            4.0       4.0       8.0      24.0      4.0      51.0
                  100.0     100.0      66.7      54.5     33.3
                    2.0       2.0       4.1      12.2      2.0


          2.        0         0         1        5         2        24
    FEMALE          0         0         4.2      20.8      8.3      49.0
                    0         0        33.3      45.5     66.7
                    0         0         2.0      10.2      4.1


         COLUMN     1         1         3       11         3        49
          TOTAL    2.0       2.0       6.1      22.4      6.1      100.0
    (CONTINUED)
```

Figure 4–5   *Continued*

```
    SEX        SEX OF SUBJECT
BY EDUC


              EDUC
      COUNT
     ROW PCT                                         ROW
     COL PCT                                        TOTAL
     TOT PCT        14.       15.      16.      17.
SEX
        1.           1         3        7        3        25
MALE               4.0       12.0     28.0     12.0      51.0
                  25.0       75.0     50.0     37.5
                   2.0        6.1     14.3      6.1

        2.           3         1        7        5        24
FEMALE            12.5        4.2     29.2     20.8      49.0
                  75.0       25.0     50.0     62.5
                   6.1        2.0     14.3     10.2

        COLUMN       4         4       14        8        49
         TOTAL     8.2        8.2     28.6     16.3     100.0

RAW CHI SQ =       5.23935 WITH      8 D.F., SIG. =    .7317
CRAMER"S V =   .32699
CONTINGENCY COEFFICIENT =   .31080
LAMBDA (ASYMMETRIC) =       .20833 WITH SEX       DEP.
                    =            0 WITH EDUC       DEP.
LAMBDA (SYMMETRIC)   =      .08475
UNCERTAINTY COEF. (ASYMMETRIC) =     .09014 WITH SEX      DEP.
                    =                .03289 WITH EDUC      DEP.
UNCERTAINTY COEF. (SYMMETRIC)   =    .04819
KENDALL"S TAU B =   .14056, SIG. =   .1352
KENDALL"S TAU C =   .17993, SIG. =   .1352
GAMMA =   .21774
SOMERS"S D (ASYMMETRIC) =    .10976 WITH SEX       DEP.
                    =        .18000 WITH EDUC       DEP.
SOMERS"S D (SYMMETRIC)   =   .13636
ETA =   .32699 WITH SEX       DEPENDENT.
ETA =   .17015 WITH EDUC      DEPENDENT.
PEARSON"S R =   .17015, SIG. =  .1212

MISSING OBSERVATIONS -        1
```

The PEARSON CORR procedure calculates all possible correlations among the variables listed. Further limitations on which correlations are calculated can be accomplished through the use of WITH. An example of the use of WITH is

PEARSON CORR;INCOME WITH EDUC,MARSAT

The WITH in this command tells SPSS to correlate only the variables listed on the left side of the word with the variables listed on the right side. This command would produce the correlations of INCOME with EDUC and INCOME with MARSAT, but it would *not* calculate EDUC with MARSAT.

An example of a printout of a PEARSON CORR run using option 6 is shown in Figure 4-6.

Figure 4–6   SPSS PEARSON CORR Example

The Program

```
10.      VARIABLE LIST
10.005 SUBNUM,SEX,INCOME,EDUC,RELIG,MARSAT
15.      INPUT FORMAT
15.005 (F3.0,1X,F1.0,1X,F5.0,1X,F2.0,1X,F1.0,1X,F1.0)
20.      N OF CASES
20.005 50
25.      VAR LABELS
25.005 SEX,SEX OF SUBJECT/RELIG,RELIGIOUS PREFERENCE
30.      VALUE LABELS
30.005 SEX(1)MALE(2)FEMALE/MARSAT(1)LOW(7)HIGH
35.      MISSING VALUES
35.005 SEX TO MARSAT(BLANK)
40.      PEARSON CORR
40.005 INCOME,EDUC,MARSAT
45.      OPTIONS
45.005 6
```

The Data File

See Box 4–1.

The Output

```
- - - PEARSON CORR - - -

00045600 CM NEEDED FOR PEARSON CORR




INCOME      .8296   INCOME      .4786   EDUC        .3591
WITH      N( 48)    WITH      N( 48)    WITH      N( 48)
EDUC       P=.000   MARSAT     P=.000   MARSAT     P=.006


A VALUE OF 99.0000 IS PRINTED IF A COEFFICIENT CANNOT BE COMPUTED.
```

SCATTERGRAM

SCATTERGRAM produces a graph that plots the data used in a Pearson correlation coefficient. The specification for SCATTERGRAM is the same as that used for PEARSON CORR. For example:

SCATTERGRAM;INCOME,MARSAT

Several options are available. Most of these affect the visual appearance of the graph. Option 4 suppresses grid lines in the graph. Six statistics are available, including a Pearson r (statistic number 1). When using SCATTERGRAM, it is easiest to specify STATISTICS;ALL.

A SCATTERGRAM output is shown in Figure 4–7.

Figure 4-7 SPSS SCATTERGRAM Example
The Program

```
10.      VARIABLE LIST
10.005 SUBNUM,SEX,INCOME,EDUC,RELIG,MARSAT
15.      INPUT FORMAT
15.005 (F3.0,1X,F1.0,1X,F5.0,1X,F2.0,1X,F1.0,1X,F1.0)
20.      N OF CASES
20.005 50
25.      VAR LABELS
25.005 SEX,SEX OF SUBJECT/RELIG,RELIGIOUS PREFERENCE
30.      VALUE LABELS
30.005 SEX(1)MALE(2)FEMALE/MARSAT(1)LOW(7)HIGH
35.      MISSING VALUES
35.005 SEX TO MARSAT(BLANK)
40.      SCATTERGRAM
40.005 INCOME,MARSAT
45.      OPTIONS
45.005 4
50.      STATISTICS
50.005 ALL
```

The Data File

See Box 4-1.   (*continued*)

T-TEST
The T-TEST procedure allows a test of whether the mean scores of two groups are significantly different from one another. There are two versions of the T-TEST command.

The first version applies to designs in which there are different subjects in each of the groups (independent group designs). For example, one group of subjects is exposed to loud noise, and another group of subjects is treated to low-level noise. Remember that a column of data must be created to indicate whether a subject was in group 1 or group 2. Suppose this variable is NOISE. Now suppose that all subjects are measured on the number of errors made while performing a task. If the name of this dependent variable is ERRORS, the T-TEST command would be

    T-TEST;GROUPS=NOISE(1,2)/VARIABLES=ERRORS

The first part of the T-TEST command specifies the name of the group variable and in parentheses tells SPSS that subjects are in either group 1 or group 2. The second part lists the dependent variables that were measured in the study.

The second version of T-TEST is used when the same subjects are used in both groups (called repeated measures or paired samples). In the noise study example, all subjects would be exposed to both high- and

Figure 4–7   *Continued*

The Output

```
- - - SCATTERGRAM - - -

GIVEN   2 VARIABLES, INITIAL CM ALLOWS FOR   1831 CASES
                     MAXIMUM CM ALLOWS FOR   8657 CASES


SCATTERGRAM OF    (DOWN)  INCOME
                  (ACROSS) MARSAT

                    1.67      3.00      4.33      5.67      7.00
          .+----+----+----+----+----+----+----+----+----+.
55000.00  +                                          *     +   55000.00
          I                                                I
          I                                                I
          I                                                I
          I                                                I
49600.00  +                                                +   49600.00
          I                                                I
          I                                                I
          I                                                I
          I                                                I
44200.00  +                                                +   44200.00
          I                                                I
          I                                    *           I
          I                                              *I
          I                                                I
38800.00  +                                                +   38800.00
          I                                                I
          I                               *                I
          I                               *        *     *I
          I                                                I
33400.00  +                                                +   33400.00
          I                                              *I
          I               *                *        *      I
          I               *                        2I
          I                                    2    *I
28000.00  +                               *            *+   28000.00
          I                                          *I
          I                                  2         I
          I               *                  *         I
          I                                            I
22600.00  +                          *              *      +   22600.00
          I          *                        3         I
          I                                  *          I
          I                                            I
          I*                                           I
17200.00  +                                  *     *      +   17200.00
          I          *                                  I
          I          *                *      *        *I
          I*                                           I
11800.00  +               *          *                  +   11800.00
          I               *          *                  I
          I               *                  *        *I
          I                          2                  I
          I                                            I
 6400.00  +          *                                   +    6400.00
          .+----+----+----+----+----+----+----+----+----+.
            1.00      2.33      3.67      5.00      6.33
```

Figure 4–7 *Continued*

```
STATISTICS..

   CORRELATION (R)-              .47860        R SQUARED       -           .22905
   SIGNIFICANCE R -              .00029        STD ERR OF EST -      9262.92633
   INTERCEPT (A)   -        9462.24478         STD ERROR OF A -      4027.98996
   SIGNIFICANCE A -              .01158        SLOPE (B)       -      2856.97564
   STD ERROR OF B -          772.80550         SIGNIFICANCE B -           .00029

   PLOTTED VALUES -      48    EXCLUDED VALUES -      0    MISSING VALUES -       2

******** IS PRINTED IF A COEFFICIENT CANNOT BE COMPUTED.
```

low-level noise and have their reaction times measured after each exposure. In this case, there would be two columns of data; one would contain reaction times under low noise, and the other would contain reaction times under high noise. Suppose these columns of data were named LOWNOISE and HINOISE. The T-TEST command to perform the analysis would be

T-TEST;PAIRS=LOWNOISE,HINOISE

The word PAIRS is used and is followed by the names of the variables containing the data. Normally no options or statistics are specified with the T-TEST procedure. An example of the output of a T-TEST run is shown in Figure 4–8. At this point, we will depart from our data file in Box 4–1. The data for this particular problem are shown in Figure 4–8 along with the output.

ONEWAY

ONEWAY performs a one-way analysis of variance. This is an extension of the T-TEST in which there are more than two groups of subjects. Suppose that in addition to high- and low-noise groups, the study included a no-noise group as well. Now there are more than two groups, and a one-way analysis of variance is appropriate to provide a test of whether the group means significantly differ. (ONEWAY is only appropriate for independent group designs in which different subjects are in each of the groups.)[6]

To extend the noise study example to ONEWAY, there would be one column of data containing group identification numbers 1, 2, and 3. If this variable is named NOISE and the dependent variable column is named ERRORS, then the ONEWAY command would be

ONEWAY;ERRORS BY NOISE(1,3)

Figure 4–8   SPSS T-TEST Example
The Program

```
10.       VARIABLE LIST
10.005 SUBNUM,NOISE,ERRORS
15.       INPUT FORMAT
15.005 (F3.0,1X,F1.0,1X,F2.0)
20.       N OF CASES
20.005 20
25.       T-TEST
25.005 GROUPS=NOISE(1,2)/VARIABLES=ERRORS
```

The Data File

```
001 1  1
002 1  2
003 1  0
004 1  1
005 1  1
006 1  2
007 1  1
008 1  2
009 1  0
010 1  2
011 2  4
012 2  5
013 2  6
014 2  3
015 2  6
016 2  8
017 2 10
018 2  9
019 2 11
020 2  8
```

The Output

```
- - - T-TEST - - -

00040200 CM NEEDED FOR T-TEST



GROUP 1 - NOISE    EQ      1.
GROUP 2 - NOISE    EQ      2.
```

VARIABLE ERRORS	NUMBER OF CASES	MEAN	STANDARD DEVIATION	STANDARD ERROR
GROUP 1	10	1.2000	.789	.249
GROUP 2	10	7.0000	2.625	.830

		POOLED VARIANCE ESTIMATE			SEPARATE VARIANCE ESTIMATE		
F VALUE	2-TAIL PROB.	T VALUE	DEGREES OF FREEDOM	2-TAIL PROB.	T VALUE	DEGREES OF FREEDOM	2-TAIL PROB.
11.07	.001	-6.69	18	.000	-6.69	10.61	.000

The dependent variable name is given first and separated from the group variable name with the word BY. Following the group variable name, the lowest and highest group numbers are given in parentheses. More than one dependent variable (up to 99 variables, in fact) may be listed on a single ONEWAY command.

Normally, no options are used with ONEWAY. The STATISTICS command usually uses the ALL convention:

STATISTICS;ALL

It is also possible to do polynomial contrasts and *post hoc* comparisons with ONEWAY. Check the SPSS manual for further details if you wish to pursue this feature. An example of ONEWAY output is shown in Figure 4–9.

ANOVA

The ANOVA command allows analysis of more complex experimental designs in which there are two or more independent variables. These are called factorial designs. As an example, suppose that the noise study (no noise, low noise, and high noise) also included a second independent variable of crowding. While being exposed to a specified noise level, subjects were also in a crowded or uncrowded room. This is called a 3 × 2 factorial design, and there are six different groups, as shown in Figure 4–10.

To perform the analysis of variance, two columns of data are needed for group identification (one for noise group and another for crowding group). Other data columns will contain the dependent variables such as the reaction-time measure. Suppose that in the data file the noise variable is named NOISE in the variable list and the crowding variable is named CROWD. The values of NOISE can be 1, 2, or

Figure 4–9   SPSS ONEWAY Example

The Program

```
10.      VARIABLE LIST
10.005 SUBNUM,NOISE,ERRORS
15.      INPUT FORMAT
15.005 (F3.0,1X,F1.0,1X,F2.0)
20.      N OF CASES
20.005 30
25.      ONEWAY
25.005 ERRORS BY NOISE(1,3)
30.      STATISTICS
30.005 ALL
```

(continued)

Figure 4–9   *Continued*

The Data File

```
001  1   1
002  1   2
003  1   0
004  1   1
005  1   1
006  1   2
007  1   1
008  1   2
009  1   0
010  1   2
011  2   4
012  2   5
013  2   6
014  2   3
015  2   6
016  2   8
017  2  10
018  2   9
019  2  11
020  2   8
021  3  10
022  3  12
023  3   9
024  3  10
025  3   9
026  3  12
027  3  13
028  3  15
029  3  13
030  3  15
```

The Output

```
- - - ONEWAY - - -

00045000 CM NEEDED FOR ONEWAY


VARIABLE  ERRORS
      BY  NOISE
```

ANALYSIS OF VARIANCE

SOURCE	D.F.	SUM OF SQ.	MEAN SQ.	F RATIO	F PROB
BETWEEN GROUPS	2	563.467	281.733	67.198	.000
WITHIN GROUPS	27	113.200	4.193		
TOTAL	29	676.667			

GROUP	COUNT	MEAN	STAND. DEV.	STAND. ERROR	MIN.	MAX.	95 P E R C E N T CONF INT FOR MEAN	
GRP 1	10	1.20	.79	.25	0	2.00	.64 TO	1.76
GRP 2	10	7.00	2.62	.83	3.00	11.00	5.12 TO	8.88
GRP 3	10	11.80	2.25	.71	9.00	15.00	10.19 TO	13.41
TOTAL	30	6.67			0	15.00		

Figure 4–9 *Continued*

```
       UNGROUPED DATA    4.83    .88              4.86 TO    8.47
 FIXED EFFECTS MODEL    2.05    .37              5.90 TO    7.43
RANDOM EFFECTS MODEL    5.31   3.06             -6.52 TO   19.85

RANDOM EFFECTS MODEL - ESTIM. OF BETWEEN COMPONENT VARIANCE      27.7541


TESTS FOR HOMOGENEITY OF VARIANCES

    COCHRANS C = MAX.VARIANCE/SUM(VARIANCES) =    .5477, P =  .158 (APPROX.)
    BARTLETT-BOX F =                             5.267, P =  .005
    MAXIMUM VARIANCE / MINIMUM VARIANCE =       11.071
```

3 whereas the values of CROWD can be 1 or 2. The ANOVA command to perform the analysis would be

ANOVA;ERRORS BY NOISE(1,3),CROWD(1,2)/

The dependent variable ERRORS is separated by the independent variables using the word BY. The lowest and highest values of each independent variable are given in parentheses. Up to five dependent measures may be analyzed with a single ANOVA command.

The options available for ANOVA are mainly of interest to the sophisticated user. The primary statistic of interest is number 3, which produces a printout of means and number of subjects. Thus for most uses of ANOVA, the only additional command is

STATISTICS;3

An example of ANOVA output is shown in Figure 4–11. Finally, ANOVA allows computation of analysis of covariance. Interested users should consult the SPSS manual for instructions.

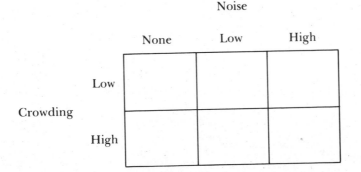

Figure 4–10 Factorial Design with Two Independent Variables

Figure 4–11 SPSS ANOVA Example

The Program

```
10.      VARIABLE LIST
10.005 SUBNUM,NOISE,CROWD,ERRORS
15.      INPUT FORMAT
15.005 (F3.0,1X,2(F1.0,1X),F2.0)
20.      N OF CASES
20.005 30
25.      ANOVA
25.005 ERRORS BY NOISE(1,3),CROWD(1,2)
30.      STATISTICS
30.005 3
```

The Data File

```
001  1  1    1
002  1  1    2
003  1  1    0
004  1  1    1
005  1  1    1
006  1  2    2
007  1  2    1
008  1  2    2
009  1  2    0
010  1  2    2
011  2  1    4
012  2  1    5
013  2  1    6
014  2  1    3
015  2  1    6
016  2  2    8
017  2  2   10
018  2  2    9
019  2  2   11
020  2  2    8
021  3  1   10
022  3  1   12
023  3  1    9
024  3  1   10
025  3  1    9
026  3  2   12
027  3  2   13
028  3  2   15
029  3  2   13
030  3  2   15
```

Figure 4–11 *Continued*

The Output

```
- - - ANOVA - - -

00050700 CM NEEDED FOR ANOVA


    CELL MEANS
      ERRORS
BY NOISE
    CROWD


TOTAL POPULATION

      6.67
  (    30)


NOISE
        1           2           3

     1.20        7.00       11.80
  (    10) (      10) (      10)


CROWD
        1           2

     5.27        8.07
  (    15) (      15)


          CROWD
            1           2
NOISE
      1      1.00        1.40
           (    5) (      5)

      2      4.80        9.20
           (    5) (      5)

      3     10.00       13.60
           (    5) (      5)


ANOVA TABLE

      ERRORS
BY NOISE
    CROWD
* * * * * * * * * * * * * * * * * * * * * * * * * * * * * * * * * * * *
```

SOURCE OF VARIATION	SUM OF SQUARES	DF	MEAN SQUARE	F	SIGNIF OF F
MAIN EFFECTS	622.267	3	207.422	155.567	.001
NOISE	563.467	2	281.733	211.300	.001
CROWD	58.800	1	58.800	44.100	.001
2-WAY INTERACTIONS	22.400	2	11.200	8.400	.002
NOISE CROWD	22.400	2	11.200	8.400	.002
EXPLAINED	644.667	5	128.933	96.700	.001
RESIDUAL	32.000	24	1.333		
TOTAL	676.667	29	23.333		

```
    30 CASES WERE PROCESSED.
     0 CASES (    0 PCT) WERE MISSING.
```

NPAR TESTS

NPAR TESTS allows computation of many nonparametric statistics. These are used in many social science investigations in which the measurements don't allow the use of statistics such as T-TEST, ONEWAY, or ANOVA. Rather than provide exact instructions for all of the tests that may be done with the NPAR TESTS procedure, the available tests are listed below. In all cases, NPAR TESTS appears on the primary line. The specification for the exact test and list of variables are written on the subline.

Test Name	SPSS Test Specification
Chi-square one sample test	CHI-SQUARE=
Kolmogorov-Smirnov	K-S=
One sample runs test	RUNS=
McNemar test for changes	MCNEMAR=
Sign test	SIGN=
Wilcoxin matched-pairs signed-rank	WILCOXIN=
Cochran Q test	COCHRAN=
Friedman 2-way ANOVA by ranks	FRIEDMAN=
Median test	MEDIAN=
Mann-Whitney U test	M-W=
Wald-Wolfowitz test	W-W=
Moses test of extreme reactions	MOSES=
Kruskal-Wallis 1-way ANOVA by ranks	K-W=

Executing the Program

Once your program is written, how is it executed? The SPSS on-line system uses a very simple command to execute the program:

    E

The symbol E (which stands for "execute") is typed without a line number and is followed by hitting the return key. SPSS will execute the program and output the results of the statistical analyses. If you are working at a CRT computer terminal, you will not automatically receive a hard-copy output. Your instructor will be able to show you the method or methods for obtaining a printout on your computer.

EDIT

A useful command to use on SPSS after the program has been written is called EDIT. EDIT is used to check the SPSS program for a variety of errors. The normal method of using EDIT is to insert the word EDIT as line 1 of the program. It looks *exactly* like this:

1 EDIT

The program is then executed using E; because EDIT is the first line of the program, no statistics are output. Rather, you will be informed of any errors detected in the program (caution: some errors may not be detected, however). The SPSS program can then be corrected if necessary. Before obtaining the statistics, the line containing EDIT must be deleted. This is done by typing

D,1

Now E can be used to execute the program. The statistical command will never be executed if EDIT is still in the program.

WRITE and READ

Preparing and typing an SPSS program can be time-consuming. Often users would like to be able to use the SPSS program on numerous occasions because all desired statistical analyses may not be accomplished on a single run. SPSS allows the user to permanently save an SPSS program as a file. This file can be saved and later changed and replaced as can any computer file (see Chapter 2).

Saving an SPSS program requires two steps. First, while still in SPSS the user types the command WRITE followed by a unique file name. For example:

WRITE,PROG1

No line number is used with the WRITE command. The use of WRITE causes a file containing the SPSS program to be placed in the computer's main memory under the designated file name. The user then exits the SPSS package with the END command. Now the user must save the program file in the computer's permanent memory system. After leaving the SPSS system, the command for saving a file on your system is used. For example:

SAVE,PROG1

After an SPSS program has been saved as a permanent file, it can be used again without having to be retyped. This is accomplished with the READ command. When you get to the point in SPSS where you would normally input a program, you can retrieve the saved program by simply typing READ followed by the file name. For example:

    READ,PROG1

If it is necessary to save a revised version of the SPSS program file, this can be done by using the WRITE command again. Immediately after leaving SPSS, you may save the revised version by using the procedure for replacing files that is employed on your system. (*Note:* Your computer may require that you employ a different procedure for saving and changing SPSS programs.)

Exiting SPSS

When finished with the SPSS system, the user must leave the SPSS program. This is done by typing END with no line number. You are now ready to use the computer for other purposes or to log off the computer.

Other SPSS Procedures

Numerous other SPSS procedures are available, most of which are listed below. Consult the SPSS manual if you need to use any of these.

Procedure	Function
BREAKDOWN	Descriptive statistics by level of one or more other variables
CANCORR	Canonical correlation
DISCRIMINANT	Discriminant function analysis
FACTOR	Factor analysis
GUTTMAN SCALE	Guttman scaling analysis
MANOVA	Multivariate analysis of variance
MULT RESPONSE	Analysis of data in which subjects give multiple responses to an item
NONPAR CORR	Rank-order correlation coefficients
PARTIAL CORR	Partial correlation coefficients
REGRESSION	Multiple regression analysis (also available: NEW REGRESSION)
RELIABILITY	Reliability of a scale
REPORT	Format tables for output of data

SPSS System Files

SPSS allows the construction of what are called *system files*. When the data file and the various data definition and data modification commands for the file have been prepared, the user can instruct the computer to save these as a system file. When the system file is saved, it may later be retrieved and it always contains all of the information on variable names, missing values, recoded variables, and computed variables. The user can then easily provide only the statistical commands that are desired for a particular run. Creating a system file is often desirable when the data file will be used on many occasions, or when many people will need to have access to the data but don't know about the actual structure of the data file. If you need to construct an SPSS system file, you will need to obtain information on how to do this from your instructor or the SPSS manual.

The commands and methods for saving and retrieving an SPSS system file on my computer are:

_____

_____

_____

_____

SPSS^x

SPSS^x is a new, enhanced version of the SPSS system. For the procedures and statistical tests discussed in this chapter, there are few changes from SPSS to SPSS^x. In general, the major changes involve improved methods of performing complex file management (e.g., using several data files in a single SPSS^x run) and improvements in the output produced for statistical tests. If you are using SPSS^x, most of the information in the previous sections is applicable. However, there are some important differences that will now be described. If you are an SPSS^x user, you will wish to cross-reference the material in the previous sections with the information that follows.

The SPSS^x Program

When using SPSS^x, the user no longer needs to worry about lining up instructions in the sixteenth column. The primary command must be in the first column, but the various instructions must follow the command by only one blank space. Any extensions of the primary command on subsequent cards or lines must be indented an arbitrary number of spaces (but may not be in the first column). This difference will be illustrated as we explore various SPSS^x commands in subsequent sections. It is thus somewhat easier to give instructions in SPSS^x than in the original SPSS system. However, the fundamental procedures of SPSS and SPSS^x are very similar.

An example SPSS^x program is shown below:

```
TITLE SURVEY
FILE HANDLE SURVEY
DATA LIST FILE=SURVEY/
    SUBNUM 1-3 SEX 5 INCOME 7-11 EDUC 13-14
    RELIG 16 MARSAT 18
MISSING VALUES SEX TO MARSAT(BLANK)
VAR LABELS SEX 'SEX OF SUBJECT'
    RELIG 'RELIGIOUS PREFERENCE'
VALUE LABELS SEX 1 'MALE' 2 'FEMALE'
CONDESCRIPTIVE INCOME EDUC MARSAT
FREQUENCIES VARIABLES=EDUC/
    HISTOGRAM/
```

TITLE

TITLE replaces the RUN NAME command. It is used to provide a header at the top of each page of output. An example is

TITLE ATTITUDE SURVEY

More complex methods of specifying a title or changing titles within a program are described in the SPSS^x manual.

FILE HANDLE

FILE HANDLE identifies the data file (or files) that will be analyzed. An example might be

FILE HANDLE SURVEY

This would inform SPSS^x that a data file named "SURVEY" will be used for analyses.

DATA LIST

The DATA LIST is required and replaces both VARIABLE LIST and INPUT FORMAT. Unless otherwise specified, DATA LIST assumes fixed-format data entry. An example corresponding to the one shown earlier in this chapter is

```
DATA LIST FILE=SURVEY/
      SUBNUM 1-3 SEX 5 INCOME 7-11 EDUC 13-14
      RELIG 16 MARSAT 18
```

Notice that the data list requires that the name of the file be specified with FILE=. A slash separates the file name from the list of variables. In the data list, each variable is specified with a variable name followed by the columns where the variable is located in the data file. Also notice the way in which blank spaces separate the variable names and their column specifications. When using DATA LIST, there is no need to worry about telling SPSS to skip columns. The blank columns are ignored.

If any variables have decimals, this is specified within the data list. For example, the following data list shows a variable called GPA with two decimal places, and a variable called TIME with one decimal place:

```
DATA LIST FILE=STUDENT/
      SUBNUM 1-3 GPA 4-6 (2) TIME 8-10 (1)
```

If there are two or more lines or cards per case, this must be specified in the data list with RECORDS=. The following would be necessary for a data file with three variables on two lines:

```
DATA LIST FILE=SURVEY RECORDS=2/
      1 SUBNUM 1-3 INCOME 4-8/
      2 AGE 1-2
```

This data list specifies that there are two lines of data for each subject. The first line contains data for SUBNUM and INCOME. The slash indicates that the second line should be read where AGE is located in the first two columns.

The DATA LIST is awkward when there are many variables or when you wish to use the TO convention to define a set of variables (e.g., V1 TO V10). Fortunately, you may use the more familiar "input format" style in a DATA LIST. This is shown in the following example:

```
DATA LIST FILE=QUEST/
      SUBNUM V1 TO V10
      (F3.0,10F2.0)
```

The second line of this data list specifies the variables (as in the VARIABLE LIST), while the third line indicates where the variables are located in the data file (as in the INPUT FORMAT).

N OF CASES
This is not necessary in SPSS^x.

VAR LABELS and VALUE LABELS

The VAR LABELS and VALUE LABELS commands are essentially unchanged. However, single quotation marks must enclose each label. SPSS[x] will accept the terms VAR LABELS and VARIABLE LABELS as equivalent.

MISSING VALUES

The MISSING VALUES specification is unchanged in SPSS[x].

RECODE and COMPUTE

The RECODE and COMPUTE commands are essentially the same in SPSS[x]. RECODE now allows the user to create a new variable with the recoded values. Thus, the original values of the variable may be left intact. The example below would recode the values of a variable called Q1 and place the recoded version of the variable into a new variable called NEWQ1:

```
RECODE Q1 (1=5) (2=4) (4=2) (5=1) INTO NEWQ1
```

SELECT IF

There are minor changes in SELECT IF. The parentheses are optional rather than required. Also, the SELECT IF in SPSS[x] allows some very complex methods of selecting cases for analysis.

Of greatest importance is that temporary selections do not use *SELECT IF. (This is true for the *COMPUTE and *RECODE commands as well.) Temporary changes are indicated by the command TEMPORARY. For example, in order to temporarily select only males for the first statistical procedure that follows, the following would be used:

```
TEMPORARY
SELECT IF (SEX EQ 1)
```

Only males would be included in the statistical analysis specified in the lines following the SELECT IF. However, subsequent analyses would not be affected by the SELECT IF because the TEMPORARY command was used.

CONDESCRIPTIVE

The CONDESCRIPTIVE command for obtaining descriptive statistics is unchanged in SPSS[x]. The following is an example:

```
CONDESCRIPTIVE INCOME EDUC MARSAT
```

This command alone will produce the mean, standard deviation, minimum, and maximum. Other statistics are possible with a STATISTICS command.

FREQUENCIES

The FREQUENCIES command will produce frequency distributions and by default prints the mean, median, standard deviation, minimum, and maximum. The following is a simple command:

FREQUENCIES VARIABLES=SEX EDUC RELIG

Note that VARIABLES= is used to specify the variables to be analyzed. Other features of FREQUENCIES include the ability to print histograms, bar graphs, scores representing various percentile values, and a variety of statistics. The following is an example of these:

```
FREQUENCIES VARIABLES=EDUC/
        HISTOGRAM/
        BARCHART/
        STATISTICS=MODE DEFAULT
```

The STATISTICS= command is new. In SPSS^x, you must indicate the names of the statistics desired. In the example, the mode plus the default statistics would be printed.

CROSSTABS

The method for specifying a CROSSTABS run is unchanged in SPSS^x. For example:

CROSSTABS TABLES=SEX BY EDUC

The output in SPSS^x is new. Only the cell totals and associated row and column totals are printed. Option 18 must be specified to print all the cell information that is automatic in SPSS. Also, option 14 has been added to print expected frequencies. The STATISTICS command is used in the same way in both SPSS and SPSS^x.

PEARSON CORR and SCATTERGRAM

There are no changes in PEARSON CORR and SCATTERGRAM.

T-TEST

SPSS^x accepts either T-TEST or TTEST as equivalent commands. Otherwise the command operates in the same way as described earlier in this chapter.

ONEWAY and ANOVA

There are no major changes in the ONEWAY and ANOVA commands. ANOVA now accepts up to ten independent variables for an analysis of variance.

NPAR TESTS

The commands NPAR TESTS and NONPAR TESTS are considered equivalent in SPSS^x.

Conclusion

This section has highlighted differences between SPSS and SPSS^x. For the types of analyses described in this book, there are few major differences. However, the SPSS^x system does incorporate a number of improvements into almost all the procedures in the package. Moreover, the SPSS^x manual contains many improvements in style and presentation. Thus, even experienced SPSS users will want to familiarize themselves with the new features and manual when they begin using SPSS^x. SPSS^x is a new system and will coexist with SPSS for a number of years. However, eventually most colleges, universities, and businesses will replace SPSS with SPSS^x.

Summary

SPSS is the Statistical Package for the Social Sciences. SPSS requires the user to construct a data file and write an SPSS program to perform the analyses. The data file consists of a series of lines on which the data provided by each subject are recorded. Although it is possible to code nonnumeric data (e.g., male versus female), it is generally desirable to represent all data numerically (e.g., 1 = male, 2 = female).

Missing data may be coded as a blank space in the data file or may be given a specific numeric code. Decimal points may be typed but can be eliminated in the data file.

Usually data from subjects are formatted so that each variable is coded in the same column for every subject. It is possible, however, to use a free-field method in which exact column formatting is not used. Blanks cannot be used for missing data when the free-field method is used.

The SPSS program tells the computer how to interpret the data file, performs any necessary modifications of the data, and carries out the statistical analyses.

The instructions in the SPSS program are called commands. The VARIABLE LIST command gives names to the variables in the data file. These names are used in all subsequent instructions in the SPSS program. The INPUT FORMAT command tells the computer where the variables on the VARIABLE LIST are located in the data file. (DATA LIST is used as an alternative to the commands VARIABLE LIST and INPUT FORMAT.) The N OF CASES command tells the computer how many subjects are included in the data file. VAR LABELS is a command that allows the user to give longer variable names to the variables in the VARIABLE LIST. VALUE LABELS allows the user to name the values of numeric values of the variables (e.g., to label the value 1 as male and 2 as female). The MISSING VALUES command tells the computer that certain values in the data file should be considered "missing" and not be included in the statistical analyses.

Several optional SPSS commands allow data to be modified or certain data to be selected for analysis. The RECODE command changes the values of a variable. The COMPUTE command creates a new variable by performing arithmetic operations on one or more existing variables. The SELECT IF command selects a subgroup of subjects for analysis. A subgroup may be selected for only one statistical analysis with the *SELECT IF command, which may be inserted at any point in the SPSS program.

After the commands necessary to describe and modify the data file have been provided, the statistical procedure commands are placed in the SPSS program. Each statistical procedure command must specify the variables to be analyzed. Usually the procedure command is followed by a selection of OPTIONS and STATISTICS that allow the user to control certain aspects of the output. The statistical procedures described in this chapter are

CONDESCRIPTIVE
FREQUENCIES
CROSSTABS
PEARSON CORR
SCATTERGRAM
T-TEST

ONEWAY
ANOVA
NPAR TESTS

When using on-line SPSS, the user must exit the SPSS package by using the END command. The E command is used to execute the SPSS program. The program may be saved by using the WRITE command. A program that has been saved may be accessed with the READ command.

Further Reading

Hull, C. H., & Nie, N. H. (1981). *SPSS update 7–9*. New York: McGraw-Hill.

Hull, C. H., & Nie, N. H. (1981). *SPSS pocket guide, Release 9*. New York: McGraw-Hill.

Klecka, W. R., Nie, N. H., & Hull, C. H. (1975). *SPSS primer*. New York: McGraw-Hill.

Nie, N. H., & Hull, C. H. (1981). *SPSS*. New York: McGraw-Hill.

Norusis, M. J. (1982). *SPSS introductory guide: Basic statistics and operations*. New York: McGraw-Hill.

The following works pertain to the SPSS[x] system:

Norusis, M. J. (1983). *SPSS[x] introductory statistics guide*. New York: McGraw-Hill.

Norusis, M. J. (1983). *SPSS[x] advanced statistics guide*. New York: McGraw-Hill.

SPSS, Inc. (1983). *SPSS[x] user's guide*. New York: McGraw-Hill.

Notes

1. SPSS and SCSS are trademarks of SPSS, Inc.
2. Alphanumeric variables are possible but usually not used for statistical analyses and are beyond the scope of this book.
3. On some computers, the procedure is to save the SPSS program in one file and then use a single command to execute the SPSS program.
4. The RUN NAME command may not print anything on some versions of on-line SPSS. In this case, it is simply a convenience to the user who prints out the SPSS program prior to running it at the terminal.

5. Another way of creating a new variable is the IF keyword. This command allows the user to specify that when certain conditions are met ("if"), then the new variable takes on a given value.

6. The RELIABILITY command allows a one-way analysis of variance for repeated measures designs. See the SPSS manual for further information.

5

Statistical Packages II:
Minitab

Minitab is an easy-to-use statistical package developed by T. A. Ryan and his colleagues at Pennsylvania State University (Ryan, Joiner, & Ryan, 1976). Minitab may be available in either on-line interactive mode with a computer terminal or for use with punched card input. Your instructor will show you what must be done to access Minitab on your computer. Only the interactive mode will be described here. It is very easy to generalize to a punched card system, however.

On my computer, Minitab is available in ☐ interactive mode, or ☐ punched card system (Check one.)
The instruction(s) to access Minitab is (are):

_____

_____

_____

Once you have accessed Minitab, it is ready to respond to commands that are input to the computer. Each command is up to four letters in length. When it executes a command, Minitab reads the line and searches for the letters that it "knows" in its command dictionary. In practice, this means that the user can type more than the short command name on the line. For example, a Minitab line using the READ command might be

    READ DATA INTO COLUMNS NAMED C1 AND C2

When Minitab encounters this line, it ignores everything but the following:

    READ C1 C2

The extended version of the command line is simply for the convenience of the user. Similarly, the command line

    SUM THE NUMBERS IN COLUMN C1

is understood by the computer as

    SUM C1

In the remainder of this chapter, the mandatory letters are in **boldface computer type**; extra text is in `regular computer type`.

The normal sequence of events when using Minitab is as follows:

1. Access Minitab.
2. Input data into one or more columns.
3. Give statistical commands.
4. Exit from Minitab.

In the next section, data input will be described. Several of the statistical commands will then be illustrated.

Data Input: The READ Command

The data for statistical analysis may be input using the **READ** command. Data may be numerical values only; any alphabetic characters are ignored by Minitab. Generally, each column will consist of the values obtained in a study on a particular variable.

As an example, the command

READ C1 C2 C3

would allow the user to then input data in three columns of a data file. Data for the first variable would be in column C1, data for the second variable would be in column C2, and so on.

Suppose that the variables studied were sex of subject, grade point average (GPA), and college major. Data for each subject would be typed on a line always in the same order of sex (C1), GPA (C2), and major (C3). After typing the data, the user would go to the next line and input the data for the next subject. Input lines for the first 10 subjects might be as follows:

```
READ C1 C2 C3
1   2.5   2
1   3.62  5
2   2.8   5
2   3.0   5
1   2.1   3
1   3.15  2
1   2.55  2
2   1.99  6
2   2.6   5
2   3.4   4
```

There are several things to note about data input. First, variables such as sex or major must have a numeric code. Thus, "male" might be coded as 1 and "female" as 2. For major, "Business" might be 4, "English" is 3, and so on. Second, the numbers on a line must be separated by one or more blank spaces. They do not have to line up in columns (this is called free-field format). Third, if numbers are expressed with decimals, the decimal point must be typed in.

When all the data lines have been typed, any other Minitab command may be given. A useful optional command after typing the data is

END OF DATA

This command will inform you of certain types of mistakes you may have made in your data input.

It is sometimes useful to print out the contents of the data file. This is done with the **PRINT** command. To print out the data file above, you would use

PRINT C1 C2 C3

An alternative way of specifying columns in any Minitab command is as follows:

PRINT C1-C3

Minitab understands that you wish the command to apply to the two columns specified and all others in between.

You may want to save the data file so that you can use it over and over again. This may be done with the command

SAVE

You will then give the data file a name. On subsequent occasions working with Minitab, you can obtain the file by typing

RETRIEVE

There are two ways to erase the contents of a column of data. The command

ERASE C1

removes the contents of column C1. Also, any time you reuse a column with a **READ** command, the previous contents of that column are automatically erased.

Finally, it is possible to correct an error you may have made in typing your data. Suppose that you typed "2.1" in column C2 for the fifth subject but meant to type "2.2." You can correct this mistake by typing

SUBSTITUTE 2.2 INTO ROW 5 OF COLUMN C2

Minitab will locate the value currently in the fifth row of the second column and substitute the new value.

Minitab Statistical Commands

In the following sections, some of the statistical commands available in Minitab will be described.

Descriptive Statistics

Several commands provide descriptive statistics such as the mean, median, and standard deviation of a set of scores.

The statistical commands in Minitab consist of a command name that instructs the computer to do a statistical operation such as sum a set of numbers. The command is followed by a specification of one or more column numbers. Minitab then performs the statistical calculations on the data located in the columns specified. Thus the command SUM C1 tells Minitab to calculate the sum of the numbers in column C1. Minitab immediately executes the command, provides the output, and waits for another command. (*Note:* If you have also used SPSS, you will notice that the immediate execution of commands is a major difference between Minitab and SPSS.) The following commands are given as examples with an explanation of their function:

Command	Function
SUM VALUES IN C1	Gives the sum of the numbers in a column
AVERAGE VALUES IN C1	Gives the mean of the numbers in a column
STANDARD DEVIATION OF C1	Gives the standard deviation of the numbers in a column
MEDIAN OF VALUES IN C1	Gives the median of the numbers in a column
DESCRIBE C1	Gives number of subjects, mean, and standard deviation for data in a column

These examples request information for only one column. You may specify more than one column in a single command. For example,

DESCRIBE **C1-C6**

would give descriptive statistics for the data in columns C1, C2, C3, C4, C5, and C6.

Correlation Coefficients: The CORR Command

A correlation coefficient (r) provides an index of the strength of relationship between two variables (see p. 61 in Chapter 4). If you wished to correlate the variables located in columns C7 and C8, you would use the following command:

CORRELATION BETWEEN **C7** AND **C8**

If more than one column is specified, Minitab calculates the correlations among all pairs of variables. Thus the command

CORR C1 C2 C3

would result in the correlations between C1 and C2, C1 and C3, and C2 and C3.

Plotting Relationships: The PLOT Command

It is frequently useful to construct a graph in which each person's score on one variable is plotted in relation to the score on a second variable. You can visualize the relationship in addition to knowing the correlation coefficient. Minitab will construct such a graph with the **PLOT** command. For example,

PLOT C7 VS **C8**

would give the desired graph with the axes labeled C7 and C8 (example output is shown in Figure 5–1). You might note that Minitab data input does not require that the numbers in columns line up exactly.

Comparing the Means of Two Groups

A statistic to compare whether the means of two groups are significantly different is the t-test. Suppose you have collected data on the number of visits to a doctor made by individuals in two different areas of a city. A t-test would allow you to determine whether the mean number of visits in the first area is significantly different from the mean in the second area.

Figure 5–1   Minitab **PLOT** Example

The following data were read into columns C1 and C2:

```
 4  10
 6  15
 7   8
 8   9
 8   7
12  10
14  15
15  13
15  15
17  14
```

The output of the PLOT C1 C2 command is as follows:

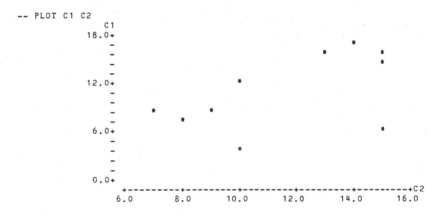

To do your analysis, the data from people in the first area would go into one column (e.g., C1) and the data from people in the second area would go into a separate column (e.g., C2). The command for the Minitab analysis is

TWOSAMPLE T-TEST FOR **C1** VS **C2**

The output is the number of subjects in each group, the mean, standard deviation, and t value.[1]

Another type of experimental design uses the same subjects in both groups. For example, in a study on the effects of crowding, subjects might be exposed to both a crowded and uncrowded environment and their reactions observed in each. This is called repeated measures or paired samples. In this case, each row of the data file would have a subject's score in the uncrowded group in one column (e.g., C10) and the same subject's score when crowded in a second column (e.g., C11).

To perform the analysis, two commands are needed:

SUBTRACT **C**10 FROM **C**11, PUT DIFFERENCES IN **C**12

TTEST **MU = 0**, FOR DIFFERENCE IN **C**12

The first command simply subtracts each score in C10 from the score in C11 and stores the difference in C12. The second command calculates a *t*-test to determine whether the difference is significantly different from zero (this is why **MU = 0** was specified).

One-Way Analysis of Variance

A one-way analysis of variance is used to test whether two *or more* group means are significantly different. The **ONEW**AY command in Minitab performs this analysis.

As an example, suppose you collect data concerning attitudes toward punishment of criminals. Now suppose you have also classified individuals in your sample into three groups: (1) those previously victimized by a crime, (2) those who have not personally been victimized but have a friend or relative who has, and (3) those who have neither been personally victimized nor have a friend or relative who has. You now want to know whether these groups differ in their attitudes.

Your data input requires that you have two columns. One column specifies the group (1, 2, or 3), and the other column contains the attitude score. If the group specification is in C1 and the attitude score in C2, the command would be

ONEWAY ANOVA ON SCORE IN **C**2, GROUPS IN **C**1

Note that the column containing the scores *must* be specified first, followed by the column containing the group identification. Minitab normally uses what is called "brief" output for analysis of variance. To obtain extended output, you must give the following command *prior* to using **ONEW**AY:

NOBRIEF

Example output using **NOBR**IEF is shown in Figure 5–2.

Two-Way Analysis of Variance

Two-way analysis of variance is the statistical procedure used with experimental designs in which the researcher simultaneously studies

Figure 5–2 Minitab **ONEW**AY Example

The following data were read into columns C3 and C4. Column C3
contains the group code variable (1, 2, or 3) and column C4 contains the
scores obtained from subjects in the three groups.

```
1  33
1  24
1  31
1  29
1  34
2  21
2  25
2  19
2  27
2  26
3  20
3  13
3  15
3  10
3  14
```

Commands and Output

```
-- NOBRIEF
-- ONEWAY ANOVA ON SCORE IN C4, GROUPS IN C3
-        NN        NLVL        NEED       IFREE       NAME      NAME1
          3           3          67          14       2524       1533

* INSUFFICIENT STORAGE SPACE FOR FREQUENCY DISPLAY

ANALYSIS OF VARIANCE

DUE TO        DF           SS        MS=SS/DF      F-RATIO
FACTOR         2        629.7         314.9         23.15
ERROR         12        163.2          13.6
TOTAL         14        792.9


LEVEL         N          MEAN       ST. DEV.
1             5         30.20         3.96
2             5         23.60         3.44
3             5         14.40         3.65

POOLED ST. DEV. =          3.69

INDIVIDUAL 95 PERCENT C. I. FOR LEVEL MEANS
(BASED ON POOLED STANDARD DEVIATION)
     +---------+---------+---------+---------+---------+---------+
1                                           I*******I*******I
2                               I*******I*******I
3            I*******I*******I
     +---------+---------+---------+---------+---------+---------+
    10.0      15.0      20.0      25.0      30.0      35.0      40.0
```

the effects of two independent variables on a dependent variable. In the simplest case, each of the independent variables has two groups (or levels). This is called a 2×2 factorial design and results in four different conditions. As an example, suppose that a researcher is interested in the effects of television violence and room temperature on aggressiveness. The researcher might have half of the subjects view a nonviolent program and the other half watch a violent program. Half of each of these groups watch in a very hot room, and the other half watch in a comfortable room. A measure of aggressiveness is then taken. This produces the following four groups: nonviolent-hot, violent-hot, nonviolent-comfortable, and violent-comfortable.

The analysis of variance allows the researcher to determine the separate effects of both program violence and room temperature on aggressiveness. It also reveals whether there is an interaction due to the unique combination of the two variables (e.g., an interaction would occur if subjects were very aggressive in the violent-hot condition but not aggressive in the others).

Minitab requires that the data file have two columns for specifying the groups that the subjects are in (one column for each of the two group variables) and another column for the data that were collected. Thus in our study one column would indicate whether each subject was in the violent or nonviolent group. Another column would indicate whether the subject was in the hot or comfortable room. A third column would contain the subject's aggressiveness score.

Assume that the group coding is in columns C1 and C2, and the aggressiveness score is in column C3. The Minitab command to do the analysis might be

TWOWAY, SCORE IN **C**3, GROUPS IN **C**1 AND **C**2

Remember that these examples show the required command in boldface. The rest is for convenience. Thus the previous command is the same as

TWOW C3 **C**1 **C**2

With this command, the measured variable must be listed first and the group variables next. Otherwise the output will be incorrect.

An example of **TWOW**AY output (using the **NOBR**IEF command first) is shown in Figure 5–3.

Figure 5–3 Minitab **TWOW**AY Example

The following data were read into columns C5, C6, and C7.
Column C5 contains the group code for the first independent
variable, C6 contains the group code for the second independent
variable, and C7 contains the score on the dependent variable.

```
1 1 75
1 1 70
1 1 69
1 1 72
1 1 68
1 2 90
1 2 95
1 2 89
1 2 85
1 2 91
2 1 85
2 1 87
2 1 83
2 1 90
2 1 89
2 2 87
2 2 94
2 2 93
2 2 89
2 2 92
```

Commands and Output

```
-- TWOWAY, SCORE IN C7, GROUPS IN C5 AND C6

ANALYSIS OF VARIANCE

DUE TO          DF          SS      MS=SS/DF
C5              1        361.25      361.25
C6              1        684.45      684.45
C5    * C6      1        281.25      281.25
ERROR          16        149.60        9.35
TOTAL          19       1476.55

CELL MEANS
ROWS ARE LEVELS OF C5      COLS ARE LEVELS OF C6
                                       ROW
                1           2         MEANS
        1     70.80       90.00       80.40
        2     86.80       91.00       88.90
COL.
MEANS         78.80       90.50       84.65

CELL STANDARD DEVIATIONS

                1           2
        1      2.77        3.61
        2      2.86        2.92

POOLED ST. DEV. =         3.06

INDIVIDUAL 95 PERCENT C. I. FOR LEVEL MEANS OF C5
(BASED ON POOLED STANDARD DEVIATION)
        +---------+---------+---------+---------+---------+---------+
1        I******I******I
2                                    I******I******I
        +---------+---------+---------+---------+---------+---------+
       78.0      81.0      84.0      87.0      90.0      93.0      96.0

INDIVIDUAL 95 PERCENT C. I. FOR LEVEL MEANS OF C6
(BASED ON POOLED STANDARD DEVIATION)
        +---------+---------+---------+---------+---------+---------+
1        I******I******I
2                                          I******I******I
        +---------+---------+---------+---------+---------+---------+
       75.0      78.0      81.0      84.0      87.0      90.0      93.0
```

Cross-Tabulated Frequency Tables

The **CONT**INGENCY TABLE command creates a cross-tabulated frequency table for two variables. It then calculates a Chi-square statistic. Suppose a researcher wishes to cross-tabulate sex of subject and major. If sex is coded in column C1 and major in C3, the command is

CONTINGENCY TABLE, ROWS IN **C**1, COLUMNS IN **C**3

This will produce a table with columns of the various major codes. The rows of the table are males and females. The cells of the table that is constructed show the number of males or females in each major. A sample printout using this command is shown in Figure 5–4.

If the various cell frequencies have already been calculated so that all that is desired is a Chi-square statistic, the frequencies may be input and followed by the **CHIS**QUARE command. For example, suppose you have calculated the following frequency table showing political preference and whether or not people voted in an election (the question for the social scientist is whether there is a significant tendency for persons with one political preference to be more likely to vote):

	Democrat	Republican	Other
Voted	222	198	36
Did not vote	175	149	30

To conduct the analysis, first input the frequencies in the table using the **READ** command. For example:

```
READ C1-C3
222  198  36
175  149  30
END OF DATA
```

Then use the **CHIS**QUARE command:

CHISQUARE FOR TABLE IN **C**1-C3

The output shows the calculated Chi-square statistic and degrees of freedom (see Figure 5–5).

Figure 5–4 Minitab **CONT**INGENCY Example

The following data were read into columns C8 and C9. Each
subject was classified as either 1 or 2 on the variable coded
in C8, and also classified as either 1 or 2 on the variable
coded in C9.

```
1  1
1  1
1  1
1  1
1  1
1  1
1  2
1  2
2  1
2  1
2  1
2  1
2  2
2  2
2  2
2  2
2  2
2  2
2  2
2  2
2  2
2  2
2  1
2  1
2  1
2  1
1  1
1  1
1  1
1  1
1  2
1  2
1  2
1  2
2  2
2  2
2  2
```

The Minitab command to construct a contingency table
from the data in C8 and C9, and the output, are as follows:

```
-- CONTINGENCY TABLE, ROWS IN C8, COLUMNS IN C9

EXPECTED FREQUENCIES ARE PRINTED BELOW OBSERVED FREQUENCIES

    ROW CLASSIFICATION - C8
    COLUMN CLASSIFICATION - C9
        I     1  I     2  I TOTALS
-------I-------I-------I-------
    1  I   11  I    7  I     18
       I   8.8I    9.2I
-------I-------I-------I-------
    2  I    8  I   13  I     21
       I  10.2I   10.8I
-------I-------I-------I-------
TOTALS I   19  I   20  I     39

TOTAL CHI SQUARE =

           .57 +   .54 +
           .49 +   .46 +

             =   2.06

DEGREES OF FREEDOM = ( 2-1) X ( 2-1) =   1
```

Figure 5–5   Minitab **CHIS**QUARE Example

The following data were read into columns C10, C11, and
C12 to form a contingency table.

```
222 198 36
175 149 30
```

The Minitab command to calculate the Chi-square statistic
and the output are as follows:

```
-- CHISQUARE FOR TABLE IN C10-C12

EXPECTED FREQUENCIES ARE PRINTED BELOW OBSERVED FREQUENCIES
       I   C10   I   C11   I   C12   I   TOTALS
-------I-------I-------I-------I-------
    1  I   222  I   198  I    36  I    456
       I  223.5I  195.3I   37.2I
-------I-------I-------I-------I-------
    2  I   175  I   149  I    30  I    354
       I  173.5I  151.7I   28.8I
-------I-------I-------I-------I-------
TOTALS I   397  I   347  I    66  I   .810

TOTAL CHI SQUARE =
           .01 +   .04 +   .04 +
           .01 +   .05 +   .05 +

               =    .19

DEGREES OF FREEDOM = ( 2-1) X ( 3-1) =   2
```

Other Minitab Commands

There are numerous other commands in the Minitab statistical package. For example, it is possible to

- Construct complex graphs using **TPLOT, MPLO**T, and **LPLO**T
- Join two columns together using **JOIN**
- Add, subtract, multiply, or divide values in columns
- Recode values in a column with **RECO**DE
- Select subjects with certain characteristics using **CHOO**SE
- Transform values to their square root, log, etc.

You should know that there is also another method of input of data
called the **SET** command. There are also many other statistical procedures available, including regression and several nonparametric tests.
Finally, Minitab has the ability to generate various types of distributions from which values may be sampled. This has proven to be very
useful for instruction in probability and statistics.

Exiting Minitab

A brief reminder: Any time you access a statistical package such as Minitab, you must use a command to leave the program before you can do any other work on the computer. In Minitab, the command is simply

STOP

Summary

The Minitab statistical package uses a series of commands. Each command is up to four letters in length and is followed by specifications for columns of data (e.g., C1 or C6). The **READ** command is used to input data into one or more columns. Columns of data will be shown to the user with the **PRINT** command and erased with the **ERASE** command, and corrections can be made with the **SUBSTITUTE** command. The following statistical procedures were described in this chapter:

Command	Function
SUM	Sum the values in column(s) specified.
AVERAGE	Give the mean of the values in column(s) specified.
STANDARD DEVIATION	Give the standard deviation of the values in column(s) specified.
MEDIAN	Give the median of the values in column(s) specified.
DESCRIBE	Give the mean, standard deviation, and number of subjects in column(s) specified.
CORR	Calculate one or more correlations between variables located in column(s) specified.
PLOT	Construct a graph plotting the values in two columns specified.
TWOSAMPLE	Calculate a t-test to compare two groups with data in the two columns specified.
TTEST MU = 0	Calculate a t-test for a difference score with the differences located in a column specified.

Command	Function
ONEWAY	Calculate a one-way analysis of variance with scores located in the first column specified and the group identification in the second column specified.
TWOWAY	Calculate a two-way analysis of variance with scores located in the first column specified and the group identification in the second and third columns specified.
CONTINGENCY TABLE	Construct a cross-tabulated frequency distribution for variables specified in two columns.
CHISQUARE	Calculate a Chi-square statistic for a table constructed in two or more columns specified.

The user signals the computer to exit from the Minitab package by means of the **STOP** command.

Further Reading

Ryan, T. A., Joiner, B. L., & Ryan, B. F. (1976). *Minitab student handbook.* North Scituate, MA: Duxbury Press.

Ryan, T. A., Joiner, B. L., & Ryan, B. F. (1980). *Minitab reference manual.* State College, PA: Pennsylvania State University.

Notes

1. A technical note: The **TWOS**AMPLE command uses a calculating formula that does not assume population variances to be equal. If population variances are equal, the following command may be used:

   POOLED TTEST FOR **C**1 VS **C**2

6

Statistical Packages III:
BMDP
and SAS

This chapter will explore two more statistical packages. The first is BMDP, and the second is SAS. Like SPSS and Minitab, both packages are capable of performing many types of analyses and have numerous features to aid researchers. All of these features cannot be described in a brief chapter. References to the manuals for BMDP and SAS are provided at the end of this chapter.

BMDP

BMDP (Biomedical Programs) is a package of computer programs developed by W. J. Dixon and his colleagues at the Health Sciences Computing Facility at UCLA.[1] Forty-two programs are described in the 1981 manual. These programs fall into the following categories:

- Data Description
- Data in Groups
- Plots and Histograms
- Frequency Tables
- Missing Values
- Regression
- Nonlinear Regression
- Analysis of Variance
- Nonparametric Statistics
- Cluster Analysis
- Multivariate Analysis
- Survival Analysis
- Time-Series Analysis

Each program is named beginning with the letter P. This is followed by a number and then a letter to indicate the type of program. Thus P1D and P2D are the first two descriptive statistics programs. Analysis of variance programs are designated with the letter V: P1V, P2V, etc. A list of many of the programs is provided in Table 6–1. Obviously, few if any researchers would have need for all of these. They are listed only for convenience, and in fact this chapter will examine only two of the programs. Finally, your campus computer center may not have all of these programs.

Table 6–1   BMDP Computer Programs

Data Description

P1D Simple Data Description

P2D Detailed Data Description, Including Frequencies

P4D Single Column Frequencies

Data in Groups: Description, t-Test and One-Way Analysis

P3D Comparison of Two Groups with t-Tests

P7D Description of Groups with Histograms and Analysis of Variance

P9D Multiway Description of Groups

Plots and Histograms

P5D Histograms and Univariate Plots

P6D Bivariate Plots (Scatterplots)

Frequency Tables

P1F Two-Way Frequency Tables—Measures of Association

P2F Two-Way Frequency Tables—Empty Cells and Departures from Independence

P3F Multiway Frequency Tables

P4F Two-Way and Multiway Frequency Tables

Missing Values—Patterns, Estimation and Correlations

P8D Missing Value Correlation

PAM Description and Estimation of Missing Data

Regression

P1R Multiple Linear Regression

P2R Stepwise Regression

P5R Polynomial Regression

P9R All Possible Subsets Regression

PLR Stepwise Logistic Regression

Nonlinear Regression

P3R Nonlinear Regression

PAR Derivative-Free Nonlinear Regression

PLR Stepwise Logistic Regression

Analysis of Variance and Covariance

P1V One-Way Analysis of Variance and Covariance

P2V Analysis of Variance and Covariance Including Repeated Measures

P3V General Mixed Model Analysis of Variance

P4V Univariate and Multivariate Analysis of Variance

P8V General Mixed Model Analysis of Variance with Equal Cell Sizes

Nonparametric Analysis

P3S Nonparametric Statistics

Cluster Analysis

P1M Cluster Analysis of Variables

P2M Cluster Analysis of Cases

P3M Block Clustering

P8M Boolean Factor Analysis

P9M Linear Scores from Preference Pairs

PKM K-Factor Clustering of Means

Multivariate Analysis

P4M Factor Analysis

P6M Canonical Correlation Analysis

P6R Partial Correlation and Multivariate Regression

P7M Stepwise Discriminant Analysis

Survival Analysis[a]

P1L Life Tables and Survival Functions

P2L Cox Models for Survival

Time-Series Analysis

P1T Spectral Analysis

P2T Interactive Box-Jenkins Analysis

---

[a]Survival analysis does *not* pertain to the problems observed among students of statistics and research design! Rather, it is a technique used in medical research.

Running a BMDP Program

BMDP requires that a data file be constructed and that an additional file contain the instructions for the BMDP program. The exact procedures for running the program will differ depending upon whether you are using an on-line interactive system or a punched card batch system, and also on the unique requirements of your computer system. On a typical punched card system, the procedure requires the following:

1. A card to call the specific BMDP program (e.g., P2V)
2. A set of cards containing the control language necessary to run the program
3. A set of cards containing the data file
4. A card telling the computer that the run is complete

Your instructor will provide more information on accessing and running BMDP programs on your computer.

BMDP on my computer system is   ☐ on-line interactive, or
☐ punched card batch (Check one.)
The instructions to access BMDP on my computer system are

_____

_____

_____

_____

_____

_____

_____

_____

_____

_____

The BMDP Data File

The procedure for constructing a BMDP data file is exactly the same as the procedure for SPSS data files described in Chapter 4. You may wish to review that material now because this chapter will not cover data file construction with the same degree of depth.

Consider the following data file containing the variables of subject identification number, sex, income, education, religious preference, marital satisfaction, and GPA.

```
001  1  10067  12  2  6  267
002  2  12962  13  3  7  381
003  2  25000      1  7  300
004  1  36079  16  1  5  196
005  1  30005  16  2  4  208
006  1  18967  11  2  2  276
007  2  19010  14  3  3  312
008  2         12  1  3
009  1   7650  12  1  4  290
010  2  24968  16  1  6  326
```

You should first note that variables are entered in exactly the same order for all subjects and that the data line up in the same columns for all subjects. In this data file, a space was left between each variable although this is not required. It is good practice, however, never to leave more than one space between variables. Also note that blanks appear in some of the columns of variables. The blanks indicate missing values; it is usually easiest to let a blank represent missing data for a subject. Now look at the values for GPA shown in the last column of the data file. In actuality, GPA values contain a decimal point; GPA can range from 0.00 to 4.00 (at most colleges). However, the decimal point was not typed in the data file. Including the decimal point is optional in BMDP data files. Finally, note that all variables are coded numerically. For example, sex of subject might be coded so that a 1 indicates a male and a 2 indicates a female.[2]

BMDP Control Language

The instructions to have the computer analyze a particular data file are referred to as control language. Certain fundamental control language instructions are common to all BMDP programs. Each program also has its own additional instructions that are required to run the particular program. An example of the fundamental instructions for a BMDP program is shown here.

```
/ PROBLEM      TITLE IS 'TEST DATA'.
/ INPUT        VARIABLES ARE 7.
               FORMAT IS '(F3.0,1X,F1.0,1X,F5.0,1X,F2.0,1X,F1.0,1X,F1.0,1X,F3.2)'.
               CASES ARE 10.
/ VARIABLE     NAMES ARE SUBNUM,SEX,INCOME,EDUC,RELIG,MARSAT.
               BLANKS ARE MISSING.
/ END
```

PROBLEM, INPUT, and VARIABLE are all thought of as "paragraphs." Each paragraph begins with a slash, and is followed by "sentences" that describe the data for the computer. An arbitrary number of spaces separates the "paragraph" title from the "sentence." Each sentence ends with a period. Each of these instructions requires further explanation. We will now examine each of the paragraphs in the program.

```
/ PROBLEM      TITLE IS 'TEST DATA'.
```

The PROBLEM instruction is required. The sentence specifying a title is optional but useful to identify the analysis. The title is enclosed by single quotation marks and is printed on the output. If the title is not desired, the word / PROBLEM by itself is acceptable.

```
/ PROBLEM
/ INPUT        VARIABLES ARE 7.
               FORMAT IS '(F3.0,1X,F1.0,1X,F5.0,1X,F2.0,1X,F1.0,1X,F1.0,1X,F3.2)'.
               CASES ARE 10.
```

The INPUT paragraph tells the computer the number of variables in the data file, the format of the variables, and the number of cases. The VARIABLES ARE 7 sentence indicates that there are seven variables in the data file.

The FORMAT statement specifies where the variables are located in the data file. The format must be enclosed within parentheses and contain the single quotation marks. F3.0 specifies that the first variable is located in the first three columns of the data file. The ".0" indicates that there are no decimals in the numbers used for the variable. 1X specifies that the computer should skip the next column because no data are located there. F1.0 indicates that the next variable occupies one column and that there are no decimal points. The remaining elements in the format continue in the same fashion. Check to make sure that you understand why the format given describes the data file shown in the previous section. The last element in the format is F3.2. This indicates that the last variable occupies three columns in the data file and that there are two decimal places. This will cause the computer to read a GPA value in the data file with the correct decimal point.[3]

The CASES ARE 10 sentence tells the computer that there are 10 subjects (cases) in the data file. This sentence is optional.

In the above examples, the words ARE and IS were used. The computer understands ARE, IS, and = as equivalent, and they may be used interchangeably.

```
/ VARIABLE    NAMES ARE SUBNUM,SEX,INCOME,EDUC,RELIG,MARSAT.
              BLANKS ARE MISSING.
```

The VARIABLE paragraph provides variable names and information on missing values. The names can be up to eight characters in length and are used primarily for clarity in the printed output. BLANKS ARE MISSING indicates that blanks are considered missing data and not included in the calculations. Other ways of coding missing values are described in the BMDP manual. If the data file contains no missing data, this sentence is, of course, not necessary.

After the required paragraphs that we have been discussing are inserted, subsequent optional paragraphs can be inserted to perform data modification or selection. This is accomplished with the TRANSFORM paragraph. Consider the following:

```
/ TRANSFORM    X=A+B.
```

This would create a new variable called X by adding the values of variables A and B. Subtraction, multiplication, division, exponentiation, and trigonometric functions may also be performed. Specific cases can also be selected or deleted with the TRANSFORM paragraph.

Any paragraphs that are required by a specific BMDP program are now inserted. The BMDP control language must end with the following paragraph:

```
/ END
```

On a punched card system, the control language instructions are followed by the cards containing the data. Again, your instructor will provide the exact instructions for your computer.

If error messages appear when you are running a BMDP program, you can usually trace them to the control language instructions. The following should routinely be checked:

1. Does each "paragraph" begin with a slash?
2. Does each "sentence" end with a period?

3. Are single quotation marks used to enclose the title and the format?
4. Are there equal numbers of left and right parentheses in the format?
5. Are all variables specified?
6. Does the format specification correspond to the way the variables were coded in the data file?

In the following sections, we will illustrate two BMDP programs and show the output from these. The programs are P6D—bivariate plots (scatterplots)—and P2V—analysis of variance and covariance including repeated measures.

P6D—Bivariate Plots (Scatterplots)

P6D produces a graph in which values of two variables are plotted to show the relationship between the variables. To run P6D, a PLOT paragraph is necessary to specify the variables to be plotted. Several scatterplots may be obtained in a single PLOT paragraph.

In the following example, suppose a researcher has obtained data from 25 students on the grades they gave their instructor on an instructor evaluation form. The program and data to obtain a scatterplot are given here. In the data file, both grades and instructor evaluations are on a 1–5 scale, a subject number is included, and there are no missing data.

```
/ PROBLEM    TITLE IS 'GRADE DATA'.
/ INPUT      VARIABLES ARE 3.
             FORMAT IS '(F3.0,2F2.0)'.
             CASES ARE 25.
/ VARIABLE   NAMES ARE SUBNUM,GRADE,EVAL.
             BLANKS ARE MISSING.
/ PLOT       YVAR IS GRADE.
             XVAR IS EVAL.
/ END
004  3  4
002  3  4
003  5  3
004  3  3
005  3  4
006  2  1
007  4  5
008  5  5
```

```
009  4  4
010  4  3
011  3  4
012  3  3
013  3  4
014  1  2
015  3  5
016  4  5
017  2  3
018  3  3
019  3  4
020  5  4
021  3  4
022  4  3
023  3  5
024  5  4
025  2  3
```

The PLOT paragraph requires that the names of the variables to be plotted on the *y* axis (vertical) and the *x* axis (horizontal) be specified. The output from this program is shown in Figure 6–1.

P2V—Analysis of Variance and Covariance
Including Repeated Measures

P2V performs an analysis of variance (or covariance) for several types of experimental designs and may be used with equal or unequal numbers of subjects in each group. It may be used for designs in which different subjects are in each group (i.e., independent groups designs), the same subjects are in all groups (i.e., repeated measures designs), or a combination of these (i.e., mixed independent groups and repeated measures designs).

The first example is a simple two-way analysis of variance. We will use the same design and data employed in Chapter 4. In that example, subjects were exposed to no noise, low noise, or high noise while in either a crowded or uncrowded room. This is a 3 × 2 factorial design with six groups. While in the particular experimental condition, subjects are measured on the number of errors made when performing a task. The number of errors is the dependent variable. The data file is exactly the same as the one used in Chapter 4. One variable codes the noise condition (1, 2, or 3), a second codes the crowding condition (1 or 2), and a third codes the dependent variable scores. A program to perform the analysis is shown here.

Figure 6–1 Example Output from BMDP Program
P6D Bivariate Plots (Scatterplots)

1BMDP6D—Bivariate (Scatter) Plots
Health Sciences Computing Facility
University of California, Los Angeles 90024
Copyright © 1977, The Regents of the University of California

```
VARIABLES TO BE USED
                  1  SUBNUM        2  GRADE         3  EVAL

  GROUPING VARIABLE . . . . . . . . . . . . . .    (NONE GIVEN)

  NUMBER OF CASES READ. . . . . . . . . . . . .       25
1
  TABLE OF CONTENTS

  HORIZONTAL      VERTICAL
  VARIABLE        VARIABLE     GROUP      PLOT                    PAGE
  NO. NAME        NO. NAME     NAME       SYMBOL                  NO.
0  3  EVAL         2  GRADE                         . . . . . . . 3
1
0         PLOT OF VARIABLE     3  EVAL    AND VARIABLE    2  GRADE
         .+....+....+....+....+....+....+....+....+....+....+....+....+....+....+.
         .
   5.0  +                                    1                2              1  +
         .
         .
   4.5  +                                                                       +
         .
         .
   4.0  +                                    2                1              2  +
         .
         .
   3.5  +                                                                       +
  G      .
  R      .
  A      .
  D      .
  E  3.0  +                                   3                7              2  +
         .
         .
         .
   2.5  +                                                                       +
         .
         .
   2.0  +  1                                 2                                  +
         .
         .
   1.5  +                                                                       +
         .
         .
         .
   1.0  +              1                                                        +
         .+....+....+....+....+....+....+....+....+....+....+....+....+....+....+.
          .90       1.5       2.1       2.7       3.3       3.9       4.5     5.1
              1.2       1.8       2.4       3.0       3.6       4.2       4.8
                                          EVAL
```

```
/ PROBLEM      TITLE IS 'BMDP EXAMPLE PROGRAM'.
/ INPUT        VARIABLES ARE 4.
               FORMAT IS '(F3.0,2F2.0,F3.0)'.
               CASES ARE 30.
/ VARIABLE     NAMES ARE SUBNUM,NOISE,CROWD,ERRORS.
               BLANKS ARE MISSING.
/ DESIGN       DEPENDENT IS ERRORS.
               GROUPING IS NOISE,CROWD.
/ END
001   1   1    1
002   1   1    2
003   1   1    0
004   1   1    1
005   1   1    1
006   1   2    2
007   1   2    1
008   1   2    2
009   1   2    0
010   1   2    2
011   2   1    4
012   2   1    5
013   2   1    6
014   2   1    3
015   2   1    6
016   2   2    8
017   2   2   10
018   2   2    9
019   2   2   11
020   2   2    8
021   3   1   10
022   3   1   12
023   3   1    9
024   3   1   10
025   3   1    9
026   3   2   12
027   3   2   13
028   3   2   15
029   3   2   13
030   3   2   15
```

The output is shown in Figure 6–2. If you are also working with SPSS, you may wish to compare the output with the SPSS output shown in Chapter 4.

Notice that P2V required a DESIGN paragraph. The first sentence (DEPENDENT IS ERRORS.) tells the computer that ERRORS is the dependent variable. The second sentence (GROUPING IS NOISE,CROWD.) tells the computer that the independent variables are noise and crowding.

In the second P2V example, we will consider a design that mixes both independent groups and repeated measures. This is a common experimental design, and BMDP provides the easiest program for this

Figure 6–2 Example Output from BMDP Program
P2V Independent Groups Analysis of Variance

1BMDP2V—Analysis of Variance and Covariance Including Repeated
Measures
Health Sciences Computing Facility
University of California, Los Angeles 90024
Copyright © 1977, The Regents of the University of California

```
VARIABLES TO BE USED
             1 SUBNUM      2 NOISE       3 CROWD       4 ERRORS
ODESIGN SPECIFICATIONS

         GROUP =   2   3
         DEPEND =   4
0                     BEFORE TRANSFORMATION                        INTERVAL RANGE
  VARIABLE       MINIMUM     MAXIMUM     MISSING   CATEGORY  CATEGORY   GREATER    LESS THAN
  NO. NAME       LIMIT       LIMIT       CODE      CODE      NAME       THAN       OR EQUAL TO
0  2   NOISE                                       1.00000   * 1.0000
                                                   2.00000   * 2.0000
                                                   3.00000   * 3.0000
0  3   CROWD                                       1.00000   * 1.0000
                                                   2.00000   * 2.0000
ONOTE--CATEGORY NAMES BEGINNING WITH * WERE GENERATED BY THE PROGRAM.

  NUMBER OF CASES READ. . . . . . . . . . . . .       30
1
OGROUP STRUCTURE

  NOISE     CROWD       COUNT
  * 1.0000   * 1.0000     5.
  * 1.0000   * 2.0000     5.
  * 2.0000   * 1.0000     5.
  * 2.0000   * 2.0000     5.
  * 3.0000   * 1.0000     5.
  * 3.0000   * 2.0000     5.

         CELL MEANS FOR   1-ST DEPENDENT VARIABLE

                                                                      MARGINAL
  NOISE   =  * 1.0000    * 1.0000    * 2.0000    * 2.0000    * 3.0000    * 3.0000
  CROWD   =  * 1.0000    * 2.0000    * 1.0000    * 2.0000    * 1.0000    * 2.0000

  ERRORS     1.00000     1.40000     4.80000     9.20000    10.00000    13.60000     6.66667

  COUNT        5           5           5           5           5           5          30

         STANDARD DEVIATIONS FOR  1-ST DEPENDENT VARIABLE

  NOISE   =  * 1.0000    * 1.0000    * 2.0000    * 2.0000    * 3.0000    * 3.0000
  CROWD   =  * 1.0000    * 2.0000    * 1.0000    * 2.0000    * 1.0000    * 2.0000

  ERRORS      .70711      .89443     1.30384     1.30384     1.22474     1.34164
1
ANALYSIS OF VARIANCE FOR   1-ST DEPENDENT VARIABLE - ERRORS

         SOURCE                      SUM OF      DEGREES OF     MEAN              F          TAIL
                                     SQUARES     FREEDOM        SQUARE                    PROBABILITY

         MEAN                      1333.33333        1       1333.33333       1000.00       0.0000
         N                          563.46667        2        281.73333        211.30       0.0000
         C                           58.80000        1         58.80000         44.10        .0000
         NC                          22.40000        2         11.20000          8.40        .0017
1        ERROR                       32.00000       24          1.33333
```

analysis among the major statistical packages. In the simplest design of this type, different subjects are randomly assigned to the various conditions of one independent variable. This is the independent groups variable. However, all subjects receive all of the treatments in the conditions of the second independent variable. This is the repeated measures variable.

Suppose that the independent groups variable is stress. Twenty subjects are randomly assigned to work on tasks while being placed under either low or high stress (10 subjects in each of the two stress groups). Now suppose that all subjects work on both an easy and a difficult task. Task difficulty is the repeated measures variable. The dependent variable is the number of errors made on the task. To perform this analysis, the data file must include a code for each subject's group (low stress = 1 and high stress = 2) and the scores on the two tasks must be coded in two separate columns. A program to perform this analysis is shown here.

```
/ PROBLEM     TITLE IS 'REPEATED MEASURES EXAMPLE'.
/ INPUT       VARIABLES ARE 4.
              FORMAT IS '(F3.0,2F2.0,F3.0)'.
              CASES ARE 20.
/ VARIABLE    NAMES ARE SUBNUM,STRESS,TASK1,TASK2.
              BLANKS ARE MISSING.
/ DESIGN      DEPENDENT ARE 3 TO 4.
              LEVEL IS 2.
              GROUPING IS STRESS.
/ END
001  1  5   7
002  1  4   7
003  1  6   6
004  1  4   6
005  1  5   8
006  1  4   8
007  1  5   7
008  1  6   7
009  1  4   4
010  1  5   7
011  2  3  10
012  2  2   8
013  2  1   9
014  2  4  12
015  2  5  11
016  2  4   9
017  2  8  15
018  2  3  10
019  2  4   9
020  2  1   8
```

In this program, the number of errors—the dependent variable— is recorded in two columns because each subject was measured twice.

Figure 6–3   Example Output from BMDP Program P2V
Analysis of Variance with Both Independent Groups
and Repeated Measures

1BMDP2V—Analysis of Variance and Covariance Including Repeated
Measures
Health Sciences Computing Facility
University of California, Los Angeles 90024
Copyright © 1977, The Regents of the University of California

```
VARIABLES TO BE USED
            1 SUBNUM       2 STRESS        3 TASK1        4 TASK2
ODESIGN SPECIFICATIONS

         GROUP  =    2
         DEPEND =    3    4
         LEVEL  =    2
0                     BEFORE TRANSFORMATION                        INTERVAL RANGE
   VARIABLE         MINIMUM     MAXIMUM    MISSING   CATEGORY  CATEGORY   GREATER      LESS THAN
   NO. NAME         LIMIT       LIMIT      CODE      CODE      NAME       THAN         OR EQUAL TO
0  2   STRESS                                        1.00000   * 1.0000
                                                     2.00000   * 2.0000
ONOTE--CATEGORY NAMES BEGINNING WITH * WERE GENERATED BY THE PROGRAM.

   NUMBER OF CASES READ. . . . . . . . . . . . .       20
1
OGROUP STRUCTURE

   STRESS        COUNT
   * 1.0000      10.
   * 2.0000      10.

          CELL MEANS FOR   1-ST DEPENDENT VARIABLE

                                        MARGINAL
       STRESS  =  * 1.0000    * 2.0000
             R
TASK1       1    4.80000      3.50000      4.15000
TASK2       2    6.70000     10.10000      8.40000

      MARGINAL     5.75000      6.80000      6.27500

      COUNT          10           10           20

   STANDARD DEVIATIONS FOR  1-ST DEPENDENT VARIABLE

       STRESS  =  * 1.0000    * 2.0000
             R
TASK1       1      .78881     2.06828
TASK2       2     1.15950     2.13177
1
ANALYSIS OF VARIANCE FOR  1-ST DEPENDENT VARIABLE - TASK1    TASK2

       SOURCE                   SUM OF      DEGREES OF     MEAN            F        TAIL
                                SQUARES     FREEDOM        SQUARE                   PROBABILITY

       MEAN                    1575.02500       1       1575.02500      335.71      .0000
       S                         11.02500       1         11.02500        2.35      .1427
   1   ERROR                     84.45000      18          4.69167

       R                        180.62500       1        180.62500      257.02      .0000
       RS                        55.22500       1         55.22500       78.58      .0000
   2   ERROR                     12.65000      18           .70278
```

These were called TASK1 and TASK2 in the VARIABLE paragraph. The DESIGN paragraph contains three sentences:

DEPENDENT ARE 3 TO 4.

This sentence tells the computer that the third and fourth variables (TASK1 and TASK2) in the data file are the scores on the dependent variable.

LEVEL IS 2.

This sentence tells the computer that there are two levels of the *repeated measures* variable (i.e., easy versus difficult task).

GROUPING IS STRESS.

This sentence tells the computer that the variable named stress is the *independent groups* variable.

The output for this program is shown in Figure 6–3.

SAS

SAS (pronounced *sass*) is the Statistical Analysis System[4]. SAS contains many useful features in terms of data treatment, the clarity of output for graphs, charts and tables, and certain business applications. It will perform all of the various data definition, data modification, and statistical analysis procedures that are possible with the other statistical packages. However, SAS is a newer system than the other packages and is not available on as many computer systems as are the SPSS, Minitab, and BMDP packages.

Running SAS Programs
Your instructor will show you the method of accessing SAS on your computer system. Use the space provided to make notes on accessing SAS.

On my computer system, SAS is ☐ on-line interactive, or
☐ punched card batch (Check one.)
The instructions to access SAS on my computer are

SAS programs consist of data definition statements, the data to be analyzed, and statistical procedure statements. Every SAS statement ends with a semicolon.

For our examples, we will use the same data used in Chapter 4 (see Box 4–1).

The DATA Statement

The first line in the SAS program is the DATA statement that informs SAS to expect to find a data set. The DATA statement takes the following form:

DATA;

Notice that the semicolon was used at the end of the statement.

It is also possible to provide a name for your data set. The word DATA is followed by a name up to eight characters in length. For example, the following would name the data set "survey":

DATA SURVEY;

The INPUT Statement

The INPUT statement tells SAS which columns each variable is located in. The INPUT statement begins with the word INPUT, followed by the variable name. A variable name must have no more than eight characters and must begin with a letter. Next, you must indicate which columns the variables occupy. To do this you specify the column number or numbers. For instance, in our example the variable SUBNUM occupies the first three columns. This would be written in the following manner:

INPUT SUBNUM 1–3;

If the value of a variable only occupies one column, simply indicate that column number. In the example, sex is indicated in column 5. Therefore, the variables SUBNUM and SEX would be defined in the INPUT statement as

INPUT SUBNUM 1–3 SEX 5;

Remember that a semicolon must appear at the end of the INPUT statement. Now consider the following statement:

INPUT SUBNUM 1–3 SEX 5 INCOME 7–11 EDUC 13–14 RELIG 16 MARSAT 18;

This statement is telling SAS that SUBNUM occupies columns 1, 2, and 3;

column 4 is blank; SEX is in column 5; column 6 is blank; INCOME is in columns 7, 8, 9, 10, and 11; column 12 is blank; EDUC occupies columns 13 and 14; column 15 is blank; RELIG is in column 16; column 17 is blank; and MARSAT is in column 18. If there are values that you do not want SAS to read, simply do not indicate them on the input statement.

If an observation occupies more than one line or card, you must tell SAS to skip to the next line or card. For example,

INPUT SUBNUM 1–3 INCOME 76–80 #2 MARSAT 1–2;

tells SAS that SUBNUM is in columns 1 to 3, INCOME is in columns 76 to 80, and MARSAT is on card number 2 in columns 1 and 2.

It is also possible to simply list your variables on the INPUT statement if (a) each value is separated by at least one blank column, (b) missing values are represented by periods, (c) numeric values have all necessary decimal points included, and (d) character variables have no more than eight characters. For example, the INPUT statement for our example could be written

INPUT SUBNUM SEX INCOME EDUC RELIG MARSAT;

if the above conditions are met.

The LABEL Statement (Optional)
If you wish to label your variables you can do so with the LABEL statement. This statement begins with the word LABEL followed by the variable name that appeared in the INPUT statement. An equals sign is then given and followed by whatever label you wish to give the variable. In the example, the LABEL statement might look like this:

LABEL SEX=SEX OF SUBJECT RELIG=RELIGIOUS PREFERENCE;

Missing Values in SAS
SAS automatically treats blanks as missing values. It also treats periods as missing values. When printing data values, SAS prints a period to indicate that a value is missing. Each SAS procedure has a way to handle missing values, and this can be found under each individual section in the SAS manual.

Data Modification and Selection
Any instructions to transform your data may be provided at this point. There are several ways that data can be transformed in SAS. First, it is possible to create new variables from those already existing. The

procedure is similar to the COMPUTE command in SPSS. Suppose we want to create a new variable called NEWVAR, which is the summation of five questionnaire items—Q1, Q2, Q3, Q4, and Q5. The following statement would accomplish this:

NEWVAR=Q1+Q2+Q3+Q4+Q5;

The arithmetic functions are identical to those in SPSS:

+ Addition
– Subtraction
* Multiplication
/ Division
** Exponentiation

Another useful data modification tool is the IF statement. The IF statement in SAS serves multiple purposes. It is possible that you only want to carry out an action for certain observations in your data set. Perhaps you want to input lines of only your female subjects listed. You would use the IF statement

IF SEX=2 THEN LIST;

The IF statement always begins with IF followed by the condition followed by THEN followed by the statement.

The IF statement in SAS can be used to serve the same purpose as RECODE in SPSS. For example, the following five lines would reverse the scoring on a questionnaire item that used a five-point scale:

IF Q1=1 THEN Q1=5;
IF Q1=2 THEN Q1=4;
IF Q1=4 THEN Q1=2;
IF Q1=5 THEN Q1=1;

It is also possible to use the IF statement in SAS to select certain observations. If we were only interested in males, we could use either of the following statements:

IF SEX =1;

or

IF SEX NE 1 THEN DELETE;

Obviously the first statement is much less cumbersome, but both would accomplish the same goal (i.e., only include male subjects in the analysis).

The CARDS Statement

The CARDS statement appears directly before your data set. It looks like this:

CARDS;

The lines of data would follow.

After the data set, statements to instruct SAS to perform statistical analyses are provided. These are called PROC statements.

SAS Statistical Procedures

The PROC (or PROCEDURE) statement is used to analyze and process the SAS data set that you have created. The PROC statement must include the word PROC and the name of the statistical procedure you want to run. The general form is

PROC procedure name;

Some PROC statements in SAS require that you provide additional information. For example, PROC ANOVA and PROC GLM require MODEL statements. Some of the procedures available in SAS will be presented below. For additional information on the procedure, consult the SAS manual.

PROC MEANS

PROC MEANS provides summary descriptive statistics on the variables in your data. It automatically provides the mean or average as well as the number of observations (N), the standard deviation (STD), the smallest value (MIN), and the largest value (MAX). If there is room on the output it will also print the standard error of the mean (STDERR), the sum (SUM), the variance (VAR), and the coefficient of variation (CV). Upon request, it will also print other statistics as well.

The VAR statement follows PROC MEANS and indicates variables for which you would like descriptive statistics. For example, the following will produce the means and all automatic statistics on the variables INCOME, EDUC, and MARSAT:

PROC MEANS;
    VAR INCOME EDUC MARSAT;

Figure 6–4 shows the program and output for the above PROC MEANS statement.

PROC CORR

The PROC CORR statement will calculate a Pearson product-moment correlation between the variables specified as well as the associated significance probabilities. The names of the variables to be correlated must be given in the VAR statement. If no VAR statement appears, then the correlation coefficients for all numeric variables in the data set will be calculated.

To obtain the correlation between EDUC and MARSAT, the following statement would be used:

PROC CORR EDUC MARSAT;

To obtain all possible correlations between INCOME, EDUC, and MARSAT, the following statement would be used:

PROC CORR INCOME EDUC MARSAT;

Figure 6–4   SAS PROC MEANS Example
The Program

```
DATA MEANS;
INPUT SUBNUM 1-3 SEX 5 INCOME 7-11 EDUC 13-14 RELIG 16
MARSAT 18;
CARDS;
001 1 15287 12 2 7
002 2 15289 13 1 4
003 1 25100 16   3
004 1       8 1 6
005 2  9500 14 3 5
006 1 28200 15 1 5
007 1 41100 17 1 7
008 1  8666  9 1 4
009 2 12617 12 1 1
010 1 15190 12 2 2
011 2 29900 16 2 3
012 2 31000 16 1 3
013 2 26500 17 3 7
014 2  9000 10 2 4
015 1 41555 17 1 6
016 1 55000 17 1 6
017 1  6400 10 4 2
018 2 29500    4 7
019 2 28000 16 1 7
020 1 34000 16 1
021 1 12000 14 1 4
022 2 10200 12 2 3
023 1 31500 16 2 5
024 1 36000 16 4 5
025 1 15000 12 4 5
026 1 11220 10 1 4
027 1 16000 13 1 2
028 2 18500 13 2 1
029 2 10000 12 2 7
030 2 20000 12 1 5
031 2 24500 14 1 5
032 2 31000 16 3 6
033 2 35500 16 1 6
034 1 21000 16 3 2
035 1  9500 12 2 3
036 1 25650 15 1 5
037 1 18200 12 1 6
038 2 29000 17 2 6
039 2 21200 16 1 6
040 2 30500 16 3 7
041 2 36000 17 2 7
042 2 30000 17 1 7
043 2 32200 15 1 7
044 2 21200 14 1 6
045 2 18760 12 2 5
046 2 36600 17 2 5
047 1 29000 15 1 6
048 1 21500 16 1 6
049 1 22900 12 2 6
050 1 26000 16 2 5
PROC MEANS;
   VAR INCOME EDUC MARSAT;
```

Figure 6-4 Continued

The Output

NOTE: THE PROCEDURE MEANS USED 0.08 SECONDS AND 370K AND PRINTED PAGE 1.
NOTE: SAS USED 370K MEMORY.

NOTE: SAS INSTITUTE INC.
SAS CIRCLE
PO BOX 8000
CARY, N.C. 27511-8000

VARIABLE	N	MEAN	STANDARD DEVIATION	MINIMUM VALUE	MAXIMUM VALUE	STD ERROR OF MEAN	SUM	VARIANCE	C.V.
INCOME	49	23723.1428571	10435.6649245	6400.0000000	55000.000000	1490.8092749	1162434.0000	108903102.42	43.989
EDUC	49	14.1632653	2.4439290	8.0000000	17.000000	0.3491327	694.0000	5.97	17.255
MARSAT	49	4.9387755	1.7369533	1.0000000	7.000000	0.2481362	242.0000	3.02	35.170

If only certain combinations of correlations are desired, then a WITH statement must also be used. For example, to produce correlations of INCOME with EDUC and INCOME with MARSAT, but *not* of EDUC with MARSAT, the following statements would be used:

```
PROC CORR;
  VAR INCOME;
  WITH EDUC MARSAT;
```

The SAS program and output for a PROC CORR statement necessary to calculate the correlations among income, education, and marital satisfaction are shown in Figure 6–5.

Figure 6–5   SAS PROC CORR Example
The Program

```
DATA CORR;
INPUT SUBNUM 1-3 SEX 5 INCOME 7-11 EDUC 13-14 RELIG 16
MARSAT 18;
CARDS;
001 1 15287 12 2 7
002 2 15289 13 1 4
003 1 25100 16   3
004 1        8 1 6
005 2  9500 14 3 5
006 1 28200 15 1 5
007 1 41100 17 1 7
008 1  8666  9 1 4
009 2 12617 12 1 1
010 1 15190 12 2 2
011 2 29900 16 2 3
012 2 31000 16 1 3
013 2 26500 17 3 7
014 2  9000 10 2 4
015 1 41555 17 1 6
016 1 55000 17 1 6
017 1  6400 10 4 2
018 2 29500    4 7
019 2 28000 16 1 7
020 1 34000 16 1
021 1 12000 14 1 4
022 2 10200 12 2 3
023 1 31500 16 2 5
024 1 36000 16 4 5
025 1 15000 12 4 5
026 1 11220 10 1 4
027 1 16000 13 1 2
028 2 18500 13 2 1
029 2 10000 12 2 7
030 2 20000 12 1 5
031 2 24500 14 1 5
032 2 31000 16 3 6
033 2 35500 16 1 6
034 1 21000 16 3 2
035 1  9500 12 2 3
036 1 25650 15 1 5
037 1 18200 12 1 6
038 2 29000 17 2 6
039 2 21200 16 1 6
040 2 30500 16 3 7
041 2 36000 17 2 7
042 2 30000 17 1 7
043 2 32200 15 1 7
044 2 21200 14 1 6
045 2 18760 12 2 5
046 2 36600 17 2 5
047 1 29000 15 1 6
048 1 21500 16 1 6
049 1 22900 12 2 6
050 1 26000 16 2 5
PROC CORR;
  VAR INCOME EDUC MARSAT;
```

Figure 6-5 *Continued*

The Output

NOTE: THE PROCEDURE CORR USED 0.10 SECONDS AND 360K AND PRINTED PAGE 1.
NOTE: SAS USED 360K MEMORY.

NOTE: SAS INSTITUTE INC.
SAS CIRCLE
PO BOX 8000
CARY, N.C. 27511-8000

VARIABLE	N	MEAN	STD DEV	SUM	MINIMUM	MAXIMUM
INCOME	49	23723.14285714	10435.65492451	1162434.000000	6400.00000000	55000.00000000
EDUC	49	14.16326531	2.44392903	694.000000	8.00000000	17.00000000
MARSAT	49	4.93877551	1.73695331	242.000000	1.00000000	7.00000000

CORRELATION COEFFICIENTS / PROB > |R| UNDER H0:RHO=0 / NUMBER OF OBSERVATIONS

	INCOME	EDUC	MARSAT
INCOME	1.00000 0.0000 49	0.82963 0.0001 48	0.47860 0.0006 48
EDUC	0.82963 0.0001 48	1.00000 0.0000 49	0.35907 0.0122 48
MARSAT	0.47860 0.0006 48	0.35907 0.0122 48	1.00000 0.0000 49

PROC TTEST

The PROC TTEST statement allows computation of a *t*-test to determine whether the mean scores of two groups are significantly different. In the example used in Chapter 4, subjects were exposed to either low or high levels of noise while working on a task. The number of errors made on the task was the dependent variable.

To perform the analysis, the data set must have one column that codes whether each subject was in group 1 or group 2. The data on number of errors must be coded in one or more separate columns. In our example, these have been labeled NOISE and ERRORS, respectively. The SAS procedure statements to perform this analysis are

```
PROC TTEST;
    CLASS NOISE;
    VAR ERRORS;
```

The CLASS NOISE statement informs SAS that the grouping code is contained in the column labeled NOISE. The VAR ERRORS statement tells SAS that the dependent variable is located in the columns labeled ERRORS.

The SAS program and output for a PROC TTEST run are shown in Figure 6–6.

Figure 6–6   SAS PROC TTEST Example

The Program

```
DATA TTEST;
INPUT SUBNUM 1-3 NOISE 5 ERRORS 7-8;
CARDS;
001 1  1
002 1  2
003 1  0
004 1  1
005 1  1
006 1  2
007 1  1
008 1  2
009 1  0
010 1  2
011 2  4
012 2  5
013 2  6
014 2  3
015 2  6
016 2  8
017 2 10
018 2  9
019 2 11
020 2  8
PROC TTEST;
    CLASS NOISE;
    VAR ERRORS;
```

Figure 6–6 Continued

The Output

NOTE: THE PROCEDURE TTEST USED 0.09 SECONDS AND 358K AND PRINTED PAGE 1.
NOTE: SAS USED 358K MEMORY.

NOTE: SAS INSTITUTE INC.
SAS CIRCLE
PO BOX 8000
CARY, N.C. 27511-8000

TTEST PROCEDURE

VARIABLE: ERRORS

NOISE	N	MEAN	STD DEV	STD ERROR	MINIMUM	MAXIMUM	VARIANCES	T	DF	PROB > |T|
1	10	1.20000000	0.78881064	0.24944383	0.00000000	2.00000000	UNEQUAL	-6.6923	10.6	0.0001
2	10	7.00000000	2.62466929	0.82999331	3.00000000	11.00000000	EQUAL	-6.6923	18.0	0.0001

FOR HO: VARIANCES ARE EQUAL, F'= 11.07 WITH 9 AND 9 DF   PROB > F' = 0.0014

PROC ANOVA

The PROC ANOVA procedure can perform analyses of all types of designs that utilize analysis of variance. The following examples will focus on one-way and two-way analysis of variance. Consult the SAS manual for descriptions of more complex designs.

The design and data used for our one-way analysis of variance example in Chapter 4 expanded on the noise experiment described above. Subjects were exposed to either no noise, low noise, or high noise. The subjects were measured on the number of errors they made while working on a task. The two variables were named NOISE and ERRORS. To perform the analysis of variance, the following statements are needed:

```
PROC ANOVA;
    CLASSES NOISE;
    MODEL ERRORS=NOISE;
```

The CLASSES statement is used to identify the grouping variable (i.e., the independent variable, NOISE). The MODEL statement is necessary to

Figure 6–7   SAS ANOVA Example

The Program

```
DATA FACT;
INPUT SUBNUM 1-3 NOISE 5 CROWD 7 ERRORS 9-10;
CARDS;
001 1 1   1
002 1 1   2
003 1 1   0
004 1 1   1
005 1 1   1
006 1 2   2
007 1 2   1
008 1 2   2
009 1 2   0
010 1 2   2
011 2 1   4
012 2 1   5
013 2 1   6
014 2 1   3
015 2 1   6
016 2 2   8
017 2 2  10
018 2 2   9
019 2 2  11
020 2 2   8
021 3 1  10
022 3 1  12
023 3 1   9
024 3 1  10
025 3 1   9
026 3 2  12
027 3 2  13
028 3 2  15
029 3 2  13
030 3 2  15
PROC ANOVA;
    CLASSES NOISE CROWD;
    MODEL ERRORS=NOISE CROWD NOISE*CROWD;
```

Figure 6-7 Continued

The Output

NOTE: THE PROCEDURE ANOVA USED 0.12 SECONDS AND 474K AND PRINTED PAGES 1 TO 2.
NOTE: SAS USED 474K MEMORY.

NOTE: SAS INSTITUTE INC.
      SAS CIRCLE
      PO BOX 8000
      CARY, N.C. 27511-8000

ANALYSIS OF VARIANCE PROCEDURE

CLASS LEVEL INFORMATION

CLASS	LEVELS	VALUES
NOISE	3	1 2 3
CROWD	2	1 2

NUMBER OF OBSERVATIONS IN DATA SET = 30

ANALYSIS OF VARIANCE PROCEDURE

DEPENDENT VARIABLE: ERRORS

SOURCE	DF	SUM OF SQUARES	MEAN SQUARE	F VALUE	PR > F	R-SQUARE	C.V.
MODEL	5	644.66666667	128.93333333	96.70	0.0001	0.952709	17.3205
ERROR	24	32.00000000	1.33333333		ROOT MSE		ERRORS MEAN
CORRECTED TOTAL	29	676.66666667			1.15470054		6.66666667

SOURCE	DF	ANOVA SS	F VALUE	PR > F
NOISE	2	563.46666667	211.30	0.0001
CROWD	1	58.80000000	44.10	0.0001
NOISE*CROWD	2	22.40000000	8.40	0.0017

say that the dependent variable ERRORS is assumed in the statistical model to be a function of the independent variable, NOISE (thus, ERRORS=NOISE).

The second analysis of variance example is a factorial design in which subjects are exposed to no noise, low noise, or high noise while in a crowded or uncrowded room. Again, the number of errors on a task is the measured (dependent) variable. The variables were named NOISE, CROWD, and ERRORS. The following PROC ANOVA statements will perform the analysis:

```
PROC ANOVA;
    CLASSES NOISE CROWD;
    MODEL ERRORS=NOISE CROWD NOISE*CROWD;
```

The CLASSES statement is again used to identify the grouping (independent) variables, NOISE and CROWD. The MODEL statement specifies that the dependent variable, ERRORS, is a function of (1) the main effect of NOISE, (2) the main effect of CROWD, and (3) the interaction between the two independent variables — NOISE*CROWD.

The program and output for the two-way analysis of variance example are shown in Figure 6-7.

Other SAS Procedures

A few of the other statistical procedures available with SAS are listed below. Consult the SAS manual for more information on these and other statistics.

SAS Procedure	Function
PROC FREQ; TABLES variables;	Frequency distributions for variables listed
PROC FREQ; TABLES variable*variable;	Cross-tabulated frequency distribution table for the two variables listed

PROC CHART;
    VBAR variable;
    HBAR variable;
	Histogram with the variable charted on the vertical or horizontal bar

PROC PLOT; PLOT variable*variable;	Scatterplot of the two variables listed

Summary

BMDP refers to the Biomedical Computer Programs developed at UCLA. BMDP consists of a large number of programs for statistical analyses of data. Unlike SPSS and Minitab, each program in the BMDP package must be run separately. However, data prepared for one BMDP program may be used with other BMDP programs.

Instructions for a BMDP program are given in BMDP control language, which consists of "paragraphs" and "sentences." Required paragraphs include

/ PROBLEM
/ INPUT
/ VARIABLE
/ END

Other paragraphs may be used for data modification and selection. Also, each BMDP program may require certain paragraphs and sentences in addition to the paragraphs above.

SAS stands for the Statistical Analysis System. Although SAS contains many useful features and also performs the same statistical analyses as other packages, it is not as widely used or as widely available at the present time.

The first step in running an SAS program is to create an SAS data set. This is done using the DATA statement. You can give the data set a name of up to eight characters. If no name is indicated, SAS will automatically name it for you. All variables must be located on the INPUT statement. The variable names must appear as well as the corresponding columns that they occupy in the data set. If certain conditions are met, only the variable names need to be on the INPUT statement. If you wish to label your variables (this is optional), you would use the LABEL statement. Missing values in SAS are always printed as a period. SAS also automatically reads blanks as missing. If you wish to create new variables or modify existing ones, this is also possible in SAS by using the feature that allows the creation of new variables and the IF statements. The CARDS statement must appear before your data lines. This tells SAS that your data will follow.

The PROC statements in SAS allow you to analyze and process the data set that you have created. In this chapter PROC MEANS, PROC FREQ, PROC CORR, PROC TTEST, and PROC ANOVA were illustrated. SAS has many other procedures that can be found in the SAS manual.

Further Reading

Dixon, W. J. (Ed.). (1981). *BMDP statistical software 1981*. Berkeley: University of California Press.

SAS Institute, Inc. (1982). *SAS introductory guide*. Cary, NC: SAS Institute, Inc.

SAS Institute, Inc. (1982). *SAS user's guide: Basics*. Cary, NC: SAS Institute, Inc.

SAS Institute, Inc. (1982). *SAS user's guide: Statistics*. Cary, NC: SAS Institute, Inc.

Notes

1. The Health Sciences Computing Facility was sponsored by NIH Special Research Resources Grant RR-3.
2. If you are going to be using BMDP for repeated measures analysis of variance, pay particular attention to the method of data input for this design, which is described in the BMDP manual. (Also see the material on P2V later in this chapter.)
3. If the data for a subject must go on two or more lines, a slash in the format is used to indicate that more data for a subject are on the next line. For example, (F2.0/ F5.0) indicates that one variable is on the first line and another variable is on the second line. See page 41 in Chapter 4 for more information on formatting data.
4. This material was prepared with the assistance of Dr. Desdemona Cardoza.

7

Computer
Applications

T hroughout this book, we have emphasized the ability of computers to perform mathematical computations in statistical analyses. This is certainly a major use of computers by behavioral researchers, but it only begins to describe the range of applications being explored today. In fact, hardly a day goes by that an article on computer applications doesn't appear in a magazine or newspaper. This chapter will examine some of those applications, including applications in education, counseling, research, and the work environment. While many innovative and exciting uses will be described, some of the potential drawbacks and limitations will be discussed as well.

Educational Applications

In Minnesota, a state agency called the Minnesota Educational Computing Consortium (MECC) offers all school districts in the state the opportunity to access a statewide computer system and purchase microcomputers at a discount price. When a school district does this, its teachers and students may use a huge library of educational software developed by MECC. These educational programs cover a variety of academic subject areas, and the MECC project has been so successful that school districts in nearby states are joining the system and other states are creating their own systems modeled after the one in Minnesota. (See Box 7–1.)

Computer-Assisted Instruction

Computer-assisted instruction (CAI) is a term that has long been used to describe educational computer applications. For many years, the dominant type of CAI program employed the technique of programmed instruction developed by psychologist B. F. Skinner. With this technique, the student reads a small amount of material that is to be learned and is then given one or more questions to answer. If the student answers the question correctly, he or she goes to the next part of the lesson. If the student answers incorrectly, the computer program may give additional information and allow the student another attempt at answering the question, or the program may require the student to repeat the previous lesson. The advantage is that the student may arrange to take the lesson at a convenient time, work at his or her own pace, and receive immediate feedback regarding errors as well as immediate opportunity to correct errors. There are advantages for the instructor as well. Typically, the computer programs are written so that a record is kept showing how many lessons the student

Box 7–1   Carnegie-Mellon to Develop Prototype Computer Network

Carnegie-Mellon University (CMU) has taken a giant step to help its students, faculty, and staff meet the challenge of the future. In October, the university signed an agreement with IBM to develop a prototype computer network designed to provide direct personal access to CMU's full information resources. The goal of the joint venture is to lay the technological foundation for personal workstations and the related data and communications services that will be available to students, faculty, and staff, whether at home, at the office, or in the laboratory. Several thousand workstations, grouped in clusters of 10 to 15, are expected to be in place by 1986.

According to Richard M. Cyert, president of CMU, the comprehensive computing environment planned by the university differs greatly from the traditional way computer facilities have been used in higher education. President Cyert says, "In 1991, we expect to have about 7500 personal workstations, each with its own powerful computer and graphics display, all interconnected through a high-speed local area network. In addition to communications among all workstations, there will be a unified data file system and a central computing facility available to all workstations. Our objective is to extend this computing system and supporting network beyond the CMU campus to the greater Pittsburgh area through cable television or telephone lines."

CMU's integrated computing environment will encompass every discipline. Students will be able to write examinations, reports, and messages on terminals and transmit them by electronic mail. Documents in libraries, research analyses, and self-study courses will be available to workstations, as will computer aids for designing solutions to engineering and science problems. These and other capabilities are part of CMU's program to enhance learning, to provide students with competitive tools and skills, and to extend the computer environment off campus.

The agreement calls for the establishment of an Information Technology Center at the university, with funds and equipment provided by IBM. Personnel from IBM and CMU will work together at the center to develop the programming for the prototype computing environment. The agreement also provides for the establishment of a consortium of universities, with one person from each university serving as primary liaison with the CMU-IBM project. Information about the integrated computing environment will be extended to the consortium at regular meetings.

Source: *Perspectives in Computing,* vol. 2, no. 4 (December 1982), p. 49. Reprinted by permission of IBM Corporation.

has attempted, the number of errors he or she has made, and sometimes the nature of the errors. This information can help the instructor direct useful individual assistance to the student. It is used as an opportunity to individualize instruction and also to allow the instructor to channel certain types of instruction to the computer and thereby free class time for other types of activities. A few examples of ways in which instructors have used CAI illustrate this:

- In a course that demands writing, the professor tries to determine early in the semester whether students have one or more specific writing problems. A student with a problem is directed to a writing assistance center where he or she may run CAI programs on a microcomputer. Programs are available to help the student with spelling, verb usage, punctuation, and so on.
- An instructor of a laboratory course in psychology requires each student to complete a CAI program on statistics during the first two weeks of the class. This requirement allows the instructor to be confident that all students have a minimal knowledge of statistics and a recent refresher on the subject. The instructor doesn't need to spend class time lecturing on statistics.
- A foreign language instructor has students complete vocabulary drills each week. The procedure reinforces material in the text and provides a useful self-assessment for the student.

You should note that computer instruction may be designed with the goal of either teaching the student new material or providing a convenient way of giving the student drill and practice in a subject. Handheld computers such as Speak & Spell™ are used for drill and practice to help teach young children spelling and math.

Learning Game Programs
A second type of educational computer program uses games as a learning tool. Such programs utilize the graphics capabilities of the computer. In some programs, the game is used simply as a reward. That is, after the student answers a certain number of questions correctly, a game appears on the screen. In other programs, the educational goals are embedded within the game itself. That is, in the process of learning how to win the game, the student is supposed to learn information or acquire new skills. For example, a game program intended to teach students how to read maps requires the student to acquire this ability in order to accumulate points. In another game, the

The Speak and Spell™ uses a voice synthesizer that gives a child a word to spell. The child then types in the letters and receives immediate feedback on whether the word was spelled correctly. (Courtesy of Texas Instruments)

student must solve a math problem in order to score a basket in a simulated game of basketball.

Simulations

A third type of educational program uses simulations. The fundamental assumption in the use of simulations is that a computer can model real-life situations and allow students to do things in the simulation that would not be possible in the real world. For example, many business schools require students to work in a simulated business environment. Teams of students must manage companies that compete with one another. The students' success depends upon management decisions that involve such considerations as hiring practices, pay scales, advertising budgets, and union contracts. Another common simulation program uses a model to forecast future events under different possible conditions. For example, a population analysis program can show students what the future age and sex distribution of a country would be in 20 years according to various possible current birthrates. (Such forecasting programs are widely used by economists and business executives as well as educators and researchers.) In psychology and other social sciences, computer simulations are frequently used to simulate well-known studies in the field.

Videodiscs and Computers

An exciting new development in the educational application of computers is the use of videodiscs that can store and instantly access a great deal of visual and audio information. The most common current use of videodiscs is to provide an alternative to written and/or film methods to instruct persons to perform complex tasks in industrial and military settings. The student can be shown correct methods in a vivid way, be given immediate testing, and also be shown the outcome of errors. Box 7–2 explores videodisc application in more detail. Despite their promise as a teaching aid, however, few videodisc software programs are currently available and the field is limited by the cost of acquiring and developing videodisc instructional systems.

Learning to Program

In addition to using the computer as a tool for teaching material in various subject areas, many educators feel that it is important to teach students how to program computers. One view is that learning computer programming helps students acquire the computer literacy that will be necessary in a world in which the uses of computer technology can be expected to become increasingly dominant in our lives. Another view is that learning to program a computer is an excellent way for students to acquire valuable thinking and problem-solving skills. To successfully write a program, the student must analyze the problem to be solved, determine the steps necessary for the solution, apply reasoning skills in order to predict how each part of the program will actually work, test the program, analyze errors, and make corrections based on feedback from the errors. In the process, the person programming the computer enjoys the experience of making the computer "obey" commands and becomes motivated to learn new skills—such as mathematical skills—that can be used in programming.

A new programming language called Logo is being taught to achieve these goals. Logo is designed to be easy for children to use. The most common way that children use Logo is to write programs

Available Fall 1982

a reviewed and tested package from CONDUIT
Registry #PSY312/Psychology

Computer Lab in Memory and Cognition

by Janice M. Keenan
and Robert A. Keller
University of Denver

The Programs

The programs in this package are available in standard BASIC or micro versions. The standard BASIC programs are easy to implement, as they do not require tachistoscopic presentation rates, single-user computers, or special display devices with cursor addressability and screen erase. Rather, they can be implemented on a campus time-sharing system, and the only special requirement is the ability to measure response time. The micro version programs require no special equipment or timers.

Neither the standard BASIC nor micro versions of the programs assume previous experience with computers, so that students need only be taught how to log on to the computer and access the programs. The programs themselves present step-by-step instructions for the experiments, complete with sample trials, and allow the student to repeat these instructions whenever necessary.

The programs are thus extremely convenient for the instructor to use. Students generally do not need to be supervised with the package, and after the first week they can run the programs outside of class, freeing classtime for more in-depth discussion of the data and theories. In addition, all but one of the programs allow for the student's individual data to be stored in a common class data file. The programs then operate on this class file to calculate summary statistics, namely means and variances. This frees the instructor from the time-consuming tasks of collecting each student's data, tabulating all the data, and calculating statistics by hand.

Computer Lab in Memory and Cognition allows students to participate as subjects in some of the most significant experiments in the fields of memory and cognition. By actually taking part in the experiments, rather than merely reading about experimental procedures and results, students gain a deeper understanding of the basic phenomena of cognitive psychology. In addition, some of the programs provide simulations of experiments based on well-known models from the literature, demonstrations of models, or both. Because these simulations and demonstrations provide results for the same tasks that the students perform, students can compare their own results with this data and better evaluate the theories behind the experiments. The programs thus provide a dynamic, hands-on approach to the paradigms and theories of cognitive psychology.

The Package

The package for Computer Lab in Memory and Cognition consists of an instructor manual, a student manual, and ten main programs. The programs are grouped into five topic areas, and each area constitutes a separate chapter in both the instructor and student manuals. Although the chapters are written so as to integrate these topics and thus provide a coherent view of the memory system, they are also sufficiently independent that an instructor can choose among them.

The five topic areas of the package include: levels of processing (2 experiments), encoding specificity (3 experiments), semantic memory (1 experiment, 2 simulations, and 1 demonstration), sentence-picture verification (1 experiment and 1 demonstration), and constructive processes in prose comprehension (3 experiments). These topics were selected to represent a broad spectrum of current work in cognitive psychology. They are also the areas in which considerable theoretical and empirical progress has been made in recent years.*

The Student Manual

The student manual for Computer Lab in Memory and Cognition is written to stand independent of a textbook. Each chapter presents a fairly complete background on the theories and research that gave rise to the original experiment. These background sections not only set the stage for the students' experiment but also show the evolution of theories and the interplay of research findings behind current theories.

Each chapter also contains a series of discussion questions. The questions encourage and guide the student to evaluate current competing theories by using the results of the class study as well as the results of the original experiment. Inferential statistics are not required for these questions; the student is simply asked to examine the means. However, variances and sample sizes are provided in the class summary

---

*Since the topic areas of this package are concerned with long-term memory and discourse processing, they complement another of CONDUIT's psychology packages, Laboratory in Cognition and Perception by Michael Levy, et al. (CONDUIT #PSY224), which focuses on perceptual and short-term memory experiments.

Commercially available simulation programs are available for instructors. The program described above is available for classes in memory and cognition. (Reproduced by permission of CONDUIT, Oakdale, Iowa)

Box 7–2 Linking Computer and Videodisc

T he beaker tips. The chemicals slosh together in a rainbow of colors. Attentive students peer at the mixture, waiting to see what potion they have come up with.

A routine chemistry experiment? It might be, except this one won't result in any smoke bubbling out of beakers, no nasty chemical vapors, not even a dirty flask. Instead, it will be carried out on a video-display screen; the students won't wear lab coats, but there will be plenty of realism.

The electronic-chemistry lab now being developed by researchers at the University of Nebraska illustrates the potential of an emerging new technology: the interactive videodisc, the union of the computer and the videodisc.

By itself, the computer can produce startling graphics. So, too, can videodiscs flash realistic images on TV screens.

But, by linking the two, even more lifelike images are being created, resulting in a variety of new information tools popping up in classrooms, industrial plants, and shopping malls. What makes this duo versatile is its ability to combine motion-picture video with sound and color—and allow people to "interact" with the video screen.

Control the "Mixing" In the case of the electronic-chemistry lab, students will tap commands into a computer keyboard and call up video-disc pictures of a chemical-filled flask. Then, by moving a "paddle" on the computer, they will control the "mixing" of the solutions. The reaction on the screen will simulate that in a laboratory. To round out the chemistry lesson, written and verbal explanations will be tossed out to the student.

Besides being an effective teaching tool, the interactive system could eventually help schools cut down on outlays for costly chemicals and diminish the potential for accidents in the lab, says Rod Daynes, director of the university's Videodisc Design-Production Group, which is producing the program. . . .

Here are some of the areas where interactive systems are emerging:

• Industrial training. The military as well as a number of auto, aerospace, and high-technology companies are using or looking at such systems to teach complex repair tasks and train sales personnel. Because the systems combine sound and text with motion-picture demonstrations, they can be an effective (if somewhat costly) alternative to tutoring workers with films or manuals.

Researchers at the Massachusetts Institute of Technology's Architecture Machine Group, for instance, are developing an electronic-repair manual under contract with the Navy. This "visual toolbox" takes

Box 7–2 *Continued*

you step by step through procedures for tackling an automobile transmission.

• Education. Interactive systems are being devised to give students "hands on" experience with equipment too expensive for schools to own, such as electron microscopes, or to simulate actual work situations. Others are being used as teaching tools in museums, libraries, and exhibition halls. At the University of Notre Dame, students will sharpen engineering skills by "running" a power plant from the classroom. They will call up videodisc pictures, diagrams, and data about the plant by touching a word or symbol on a pressure-sensitive screen. It will also give out detailed explanations and quiz the students on plant operations.

Source: Scott Armstrong, "Linking Computer and Videodisc," Christian Science Monitor News Service, *Los Angeles Times*, March 11, 1983, Part V, p. 28.

that cause an object on a video screen (called a turtle) to draw pictures—this has been called "turtle graphics." When children play with the turtle to make it go in various directions and create desired designs, they learn about programming, math, and geometry. Perhaps the most interesting aspect of Logo is that the language is capable of doing much more than drawing pictures with the turtle. Children who are excited by their success with the turtle programs can go on to more advanced applications.

Issues in Educational Applications

The enthusiastic hope that computers will enhance the learning process must be tempered by consideration of several issues. Foremost is the issue of educational software. Hundreds of programs are available for purchase, but there is great concern about their quality. Many programs simply do not do what they claim, contain errors, and provide little useful documentation to help the user. A great deal of money is spent writing the programs, but little attention is paid to evaluating them before they are sold. This situation must change (and probably will as competition among software producers increases). In addition to the issue of software quality, there is another issue of copying the programs. Most programs purchased for microcomputers are "copy protected" and cannot be copied or changed in any

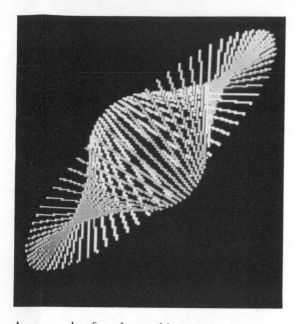

An example of turtle graphics; a program written in Logo created this design. (Courtesy of Apple Computer, Inc.)

way.[1] Copy protection prevents users from selling or giving away programs that should be purchased from the software manufacturer. However, it also prevents the user from having back-up copies in case the original diskette is damaged or from making changes in the program that the user might find desirable. A solution to this problem remains to be found.

Another issue concerns teacher training. In many schools, a micro-computer is purchased along with some software and placed in a central location. Commonly, a student gets to work on the computer as a reward. If the student does something right in the classroom, he or she is allowed to go to the computer room but receives little or no direction because the teacher has had no training in the use of computers. Even when teachers receive training, it is often minimal. For example, in a brief training session on Logo, teachers may learn how to make the turtle draw a square and a triangle. When they show students Logo in the classroom, the square and triangle are drawn and there is little chance that students will explore any further aspects of the programming language. Fortunately, teacher education programs are now beginning to provide instruction in computer technology.

Another issue concerns the nature of the games used in many educational programs. Some observers argue that the sports and battle metaphors used in many of the games appeal primarily to males. Will this fact lead to greater disparity between males and females in terms of their interest and expertise in computers and science?

There is also the issue of social equality. A concern is that wealthier school districts will purchase computers but that poorer districts will not be able to afford them. Will this create greater class disparities in a society in which computer literacy is becoming increasingly necessary? Further research on these issues is needed.

Applications in Testing and Counseling

Computers are being used in clinical and counseling psychology. One of the first applications was computer test scoring. For example, a college student might take a vocational interest test (such as the Strong-Campbell Interest Inventory) and the answer form would be mailed for computer analysis. The computer would score the test and provide detailed feedback on the results. Figure 7–1 shows one such printout. As you can see, the information provided is very extensive. It will be extremely useful in helping a counselor discuss career options with the student.

With advancements in computer technology, the entire testing process can now be conducted entirely on a microcomputer. The questions are shown on the screen and answered by the test taker at the terminal, and then the computer program provides scoring and analysis. The psychologist is saved the time it would have taken him or her to administer the test, score it, compute percentiles and norms, and so on.

The primary reason why computers will become increasingly useful for counseling and clinical diagnosis is that they can store and immediately retrieve a great deal of information about a test. Thus they can help the tester remember many things about the test, some of which he or she might overlook in a routine test interpretation. In this way, computers become valuable aids in improving the quality of the services provided to persons taking the test.[2]

Another interesting development in counseling applications is SIGI. SIGI (pronounced "siggy") is a computer-based System of Interactive Guidance and Information. It has been widely implemented in the career counseling centers in colleges and universities. The SIGI program is designed to pose questions and problems to the student in

Figure 7–1 Scored Answer Form for the Strong-Campbell Interest Inventory *(Reproduced by permission of Consulting Psychologists Press, Inc., Palo Alto, Calif.)*

SCII Profile for   SCII SAMPLE   Sex F   Age 17   Date Scored 10/26/82   Date Administered 10/11/82

Page Two — CONSULTING PSYCHOLOGISTS PRESS

Sheet I.D.: I 13 007 TWO 000558589

COUNSELOR'S COPY

Occupational Scales

Code	Scale	Sex Norm	Standard Score
IRC	Computer Programmer	M	42
IRE	Chiropractor	F	4
IRE	Chiropractor	M	17
IE	Pharmacist	M	16
I	Pharmacist	F	28
I	Biologist	F	<0
I	Biologist	M	13
I	Geographer	F	19
I	Geographer	M	21
I	Mathematician	F	23
I	Mathematician	M	16
IA	College Professor	F	20
IA	College Professor	M	28
IA	Sociologist	F	<0
IA	Sociologist	M	4
IAS	Psychologist	F	<0
IAS	Psychologist	M	11
AIR	Architect	F	0
AIR	Architect	M	7
AI	Lawyer	F	2
AI	Lawyer	M	10
AE	Public Relations Director	F	3
AE	Public Relations Director	M	17
AE	Advertising Executive	F	24
AE	Advertising Executive	M	14
AE	Interior Decorator	F	<0
AE	Interior Decorator	M	30
A	Musician	F	5
A	Musician	M	24
A	Commercial Artist	F	<0
A	Commercial Artist	M	5
A	Fine Artist	F	1
A	Fine Artist	M	4
A	Art Teacher	F	<0
A	Art Teacher	M	6
A	Photographer	F	13
A	Photographer	M	15
A	Librarian	F	23
A	Librarian	M	29
A	Foreign Language Teacher	F	18
A	Foreign Language Teacher	M	35
A	Reporter	F	<0
A	Reporter	M	24
A	English Teacher	F	19
AS	English Teacher	M	19
SA	Speech Pathologist	F	11
SA	Speech Pathologist	M	12
SA	Social Worker	F	2
SA	Social Worker	M	17
SA	Minister	F	<0
SIE	Minister	M	9
SI	Registered Nurse	F	5
S	Licensed Pratical Nurse	M	44
S	Special Ed Teacher	F	38
S	Special Ed Teacher	M	39
S	Elementary Teacher	F	44
S	Elementary Teacher	M	40
SR	Physical Ed Teacher	F	34

Occupational Scales

Code	Scale	Sex Norm	Standard Score
SR	Physical Ed Teacher	M	21
SRE	Recreation Leader	F	16
SRE	Recreation Leader	M	28
SE	YWCA Director	F	29
SE	YMCA Director	M	29
SE	School Administrator	F	23
SE	School Administrator	M	29
SCE	Guidance Counselor	M	33
SEC	Guidance Counselor	F	13
SEC	Social Science Teacher	F	27
SEC	Social Science Teacher	M	27
EA	Flight Attendant	F	13
EA	Flight Attendant	M	32
EA	Beautician	M	32
E	Beautician	F	50
E	Dept. Store Manager	F	31
E	Dept. Store Manager	M	42
E	Realtor	F	15
E	Realtor	M	41
E	Life Insurance Agent	F	16
E	Life Insurance Agent	M	13
E	Elected Public Official	F	11
E	Elected Public Official	M	23
E	Public Administrator	F	12
EI	Investment Fund Manager	M	36
EI	Marketing Executive	F	24
EI	Marketing Executive	M	33
E	Personnel Director	F	26
E	Personnel Director	M	24
E	Chamber of Commerce Exec	M	24
E	Restaurant Manager	M	48
EC	Restaurant Manager	F	32
EC	Chamber of Commerce Exec	F	41
EC	Buyer	F	29
EC	Buyer	M	35
EC	Purchasing Agent	F	30
EC	Purchasing Agent	M	31
ERC	Agribusiness Manager	M	43
ES	Home Economics Teacher	F	30
ECS	Nursing Home Administrator	M	31
EC	Nursing Home Administrator	F	35
EC	Dietician	F	16
ECR	Dietician	M	44
CER	Executive Housekeeper	F	35
CER	Executive Housekeeper	M	39
CES	Business Ed Teacher	F	54
CES	Business Ed Teacher	M	54
CE	Banker	F	60
CE	Banker	M	46
CE	Credit Manager	F	55
CE	Credit Manager	M	50
CE	IRS Agent	F	41
CE	IRS Agent	M	43
CA	Public Administrator	M	10
C	Accountant	F	43
C	Accountant	M	55
C	Secretary	F	66
C	Dental Assistant	F	50

Scale markings: 12   21   27   39   45   54   60

TIME EFFECTIVE ASSESSMENT AT LAST!

I seem to never have enough time.

True or False?

cpd

We know that obtaining med/psych histories and system reviews, and administering psychological tests is time consuming. Consequently, these assessment instruments are under-utilized and many professionals are not profiting from valuable information available to them. Now there is a thorough and efficient way to obtain this information *in your office.*

If you'd like to join the clinicians who have discovered the benefits of time effective, computerized assessment, here's what you can expect:

- More time for patient interaction
- Clinician time effectiveness
- Patient cost effectiveness
- Improved patient care
- Dual language (Spanish/English) capability serving a broad-based practice
- Large variety of sophisticated, professionally recognized tests

* Tests available:

MMPI—Minnesota Multiphasic Personality Inventory
CPI—California Psychological Inventory
MHS—Harvard Medical History Survey
JAS—Jenkins Activity Survey
PIC—Personality Inventory for Children
CAS—The Career Assessment System
SDS—Self-directed Search
VPI—Vocational Preference Inventory
ROR—Rorschach

MCDI—Minnesota Child Development Inventory
VST—Visual Searching Task
SHX—Social History
ISB—Intellectual Screening Battery
SCL-90—Symptom Check List
BECK—Beck Depression and Hopelessness Scale
ISP—Index of Somatic Problems
QI—Dissimulation Index
IER—Integrated Evaluation Report

For more information call   714/833-7931

cpd **Computerized Psychological Diagnostics, Inc.**
1101 Dove Street, Suite 200, Newport Beach, CA 92660

CPD...the professional's testing alternative

© 1982 Computerized Psychological Diagnostics, Inc.   *Psych Systems, Inc.™

A number of firms have developed programs to administer and score psychological tests that are used in clinical practice. (Reproduced by permission of Computerized Psychological Diagnostics, Inc.)

order to help the student define his or her values, explore different decision-making strategies, and discover career options. The SIGI program is completed during several sessions at a computer terminal and is followed by a discussion of the experience of the user with a career counselor. Students who have used SIGI report that it provides an enjoyable and a profitable learning experience.

Applications in
Various Fields of Research

Computers are also being used increasingly in various fields of research. On-line data collection, archival research, and literature searches are but three of the research tools made possible by computers. We will examine each of these in turn.

On-Line Data Collection
Computers are being used in the actual conduct of research in the behavioral sciences. Data can be recorded directly on a computer, and statistical analyses may be performed quickly and easily. The accuracy of the data collection and analysis may frequently increase as well.

One very useful application involves data collection by researchers in field settings. Examples would include a public opinion pollster, an anthropologist making observations in an isolated locale, and a psychologist observing children's interactions in a classroom. Small computers allow researchers to enter data that is stored and later transferred to a larger computer for analysis. The public opinion pollster uses a phone connection to send data to a central computer at the end of the day, and results are instantly analyzed. The anthropologist gets rapid feedback on the pattern of emerging data and may be able to use this to make improvements in the project that wouldn't be possible if the data had to be transported back to a university prior to analysis. The psychologist's data are recorded more quickly and accurately when he or she uses a small computer keyboard instead of making marks on a pad of paper.

Computers have also moved into research laboratories. Computers can be programmed to present instructions to subjects, display stimuli, collect response data, and then calculate statistical results. For example, Dan Kee at California State University, Fullerton, uses a microcomputer to investigate left versus right brain hemispheric dominance. Subjects are required to do a finger-tapping task while simul-

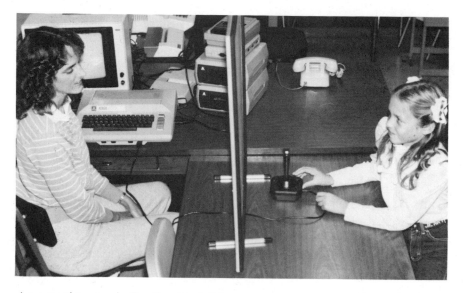

An experimenter in Dr. Dan Kee's laboratory tests a young subject on a finger-tapping task. The Atari computer is used in all phases of this research. (Photo courtesy of Dr. Daniel Kee and Kathleen Brown)

taneously engaging in some other task. The purpose is to examine interference with tapping speed while performing different tasks. The computer controls the signal to begin and end the tapping, keeps track of the number of taps in each time interval, and stores and analyzes this information. Dr. Kee's research illustrates the utility of computers in keeping track of time information. Many researchers use computers for studying reaction time following a question or a particular stimulus. Most computers may be programmed to keep time in terms of milliseconds or microseconds.

Computers are being used in animal as well as human research. A computer can be programmed to control a laboratory environment and record an animal's responses.

The technology of computerized information gathering has commercial as well as research applications. People can be linked via television to services that request attitudinal information or allow purchases of products. The possibilities for invasion of privacy in this regard are explored in Box 7–3.

Box 7–3   Two-Way Cable TV and Electronic Snooping

onsider the following situation:

C   A priest and a former nun are being interviewed on a cable-TV talk
show in Columbus, Ohio, on the issue "What is it like to be a homosexual
in Columbus?" The interviewer notes than an estimated 80,000 homosexuals live in the Columbus metropolitan area, which has an overall population of more than one million people. The interviewer then asks television
viewers the following question: How many viewers have a friend, relative,
or acquaintance who is homosexual? "If you do know a homosexual," the
interviewer says, "press button No. 1 on your home console for yes; if you
do not, press button No. 2 for no." Viewers at home use the console next
to the channel selector button on their televisions to answer the question.
Within seconds, a computer supplies the results, which are presented on
viewers' television screens: yes—65 percent; no—35 percent.

The marriage of information technology and television has created
two-way interactive cable TV. One-way cable TV carries signals, such as
first-run movies, directly into the subscriber's home. Two-way interactive
TV allows the viewer to transmit messages back to the sender through the
cable system. Experts predict that two-way TV will offer home-based
entertainment, classroomlike instruction, bill paying, shopping, home
security, medical and disaster emergency preparedness, and political and
community participation, and will attract 50 percent of American households to cable TV during the 1980s. What will be done with all the
information subscribers send over their cable systems? Will cable operators treat information provided by their customers as privileged, or will
the information be sold or passed along to confidential clients such as
mail-order houses, credit-rating companies, the Internal Revenue Service, and so on?

Currently, only a few communities in North America have two-way
TV. The best-known system operates in Columbus, Ohio. The heart of
the system is called QUBE (a trade name that doesn't stand for anything).
It is a computer that scans subscribers' households every six seconds to
find out if the TV is switched on, to what channel, and what was the last
message transmitted by the viewer to the computer. Messages from the
viewer to the computer are punched on a five-button console. The first
two buttons are used to indicate yes and no. The five buttons can be used
to answer multiple-choice questions or to punch up number codes, to
select, for instance, a product displayed on the television screen.

Two-way cable TV holds great promise. The QUBE system has, for
instance, significantly increased public participation in town meetings
and public hearings. Two public meetings on an urban redevelopment
plan in the Columbus area drew about 125 people each. When the

Box 7–3   Continued

meetings were held on two-way cable TV, the computer reported that 2000 residents participated in the two-and-a-half-hour town meetings. The questions remain, however: What happens to the information that is collected? Can the computer collate a subscriber's responses, merge them with other information on file about the subscriber (including information such as address, income, social background, and so forth), and create a file or dossier on the viewer's household?

At present neither the U.S. Congress nor state legislatures have enacted restrictive laws, nor has the industry developed policies to protect personal disclosures derived from cable TV. Experts feel that controls are coming. Until then, protection will be a matter of the cable TV operators' own policies (QUBE's operators restrict access to the computer and its records). Some local communities have restrictions to protect subscribers; other communities do not.

Psychologically, subscribers regard two-way TV as "fun." As one subscriber put it, "It gives you a sense of power, a sense of directing something far away." For instance, an audience participation program similar to the "Gong Show" is presented by QUBE. Viewers direct the show by pressing yes or no to say whether an amateur act should continue. If a majority of the viewing audience presses no, the act is ended in midperformance. On the other hand, viewers seem only vaguely aware that personal information may be tabulated and passed along to others. As one young woman said about the QUBE system, "I don't feel that I have any reason to be afraid—I may be naive, but I don't care if my opinions are recorded."

Cable-TV operators could, however, profit from the sale of information about their subscribers. For instance, the following uses (none of which are illegal) could be made of information you might provide on two-way cable TV:

1. After your recent vote during a televised town meeting about busing children to integrate public schools, you begin to receive mail and phone calls from fund-raising groups, charities, and political groups that ask for your support.
2. After watching an X-rated film, a publisher of a "skin" magazine is notified of your selection. The publisher sends you a sales brochure in a plain manila envelope.
3. A subpoena has been prepared for you. To make sure you are home to receive it, the cable operator has been paid to monitor when you turn on the TV and to report this immediately.
4. You use your two-way TV to order on credit a product advertised on TV. The store's credit manager reports that your credit pay-

Box 7–3 *Continued*

ments are in arrears. You cancel the order until you are able to transfer sufficient funds from your bank account to the store's account. However, a national credit-rating company, a confidential client of the cable operator, learns that your order has been rejected. This information is put into your dossier for the next customer who purchases your credit information.

The ultimate abuse of individual privacy by two-way TV was portrayed in George Orwell's novel *1984*. Households were under constant TV surveillance. Less chilling, but still potentially dangerous, are the abuses of files that could be obtained from two-way cable-TV subscribers. So long as businesses and government rely on personal information about individuals, there is a possibility of such abuses. It is to be hoped that the restraint of cable operators and active consumer and community monitoring will be sufficient to protect subscribers' disclosures. Industrywide regulation by cable operators, legislative action, and local ordinances can also address the potential social problem.

Source: "Loneliness and Intimate Communication" by V. J. Derlega and S. J. Margulis. In D. Perlman and P. C. Cozby (Eds.), *Social Psychology*. New York: Holt, Rinehart & Winston, 1983. Reprinted by permission.

Archival Research Data

Data collected as part of public opinion surveys and public record keeping are now routinely kept as computer data files. These data files are accessible to researchers who can conduct analyses of the data. They are termed archival research data because they are stored as archives of information available to researchers now and also in years to come. Generally these are very large files consisting of a great deal of information collected from a very large number of individuals. Examples include:

- The census taken by the United States Census Bureau every 10 years
- National surveys of political behavior and attitudes undertaken by the Survey Research Center at the University of Michigan
- Public opinion polls taken by national and regional polling firms such as the Gallup Poll
- Public health statistics maintained by government agencies
- Crime statistics maintained by government agencies

- Test results from educational tests such as the Scholastic Aptitude Test that are administered nationally each year
- Information on the attitudes and characteristics of college freshmen obtained through a cooperative arrangement of colleges and universities throughout the United States
- Survey data obtained by researchers and polling organizations in other countries

The data files may be stored on tape or disk. Researchers are able to use the files by either obtaining copies to install on a local computer or via a phone connection to a computer where the data files are located. The data can then be used to address research topics such as trends in attitudes and behavior over time; comparisons of various religious, racial, and social class groups; cross-national comparisons; and political processes. The following sections provide two examples of research projects involving archival data.

The General Social Survey

During the 1970s, the National Science Foundation provided funding for a series of surveys of adults in the United States. The project was called the National Data Program for the Social Sciences, and it was intended specifically as a resource for social scientists to use in whatever way they desire. The General Social Survey included over 200 questions that covered the following topics:

- Family and ethnic background
- Marital history and number of children
- Occupation, income, social class, job satisfaction, and occupational prestige for self and spouse
- Attitudes regarding federal spending
- Attitudes regarding international relations
- Attitudes regarding rights of antireligious persons, socialists, communists, and homosexuals
- Feelings about the most important attributes that a child should possess
- Attitudes regarding social issues such as marijuana, the death penalty, gun control, and the court system
- Experience with violence as a child or adult
- Gun ownership
- Police record
- Involvement in demonstrations

- Opinions regarding violence (such as approval or disapproval of a person's hitting a disorderly drunk)
- Confidence in institutions such as education and the press
- Trust in other people
- Attitudes regarding race relations
- Satisfaction with life
- Health
- Attitudes regarding abortion, sexual relations, and sexual materials
- Political party identification and recent voting behavior
- Religious background
- Education of self, spouse, and parents
- Union activity
- Age, race, and sex

Such a rich base of information has provided the data for numerous investigations of social behavior. The fact that this information is easily available on computers is a tremendous asset for researchers.

I.Q., Birth Order, and Family Size

A number of years ago, Lillian Belmont and Francis A. Morolla had the opportunity to study a data base consisting of the intelligence test scores of all 19-year-old men in the Netherlands. They discovered a very interesting pattern of results in the data: I.Q. is systematically related to both birth order and family size. I.Q. is higher in families with fewer children; I.Q. is higher among early-born than later-born children. Later, psychologist Robert Zajonc developed a mathematical model to explain the data, a model that is based upon the amount of intellectual stimulation received by children of differing family size and birth order. Zajonc was also able to replicate the original findings by studying a data base consisting of test scores obtained in the United States.

Computerized Literature Searches

Researchers and students in many areas of study have benefited from computer literature searches. A computer literature search provides a listing of the titles and summaries of research reports available on a particular subject. In order to perform such searches, a data base consisting of many research reports must be stored in a computer. Each report is coded by means of certain descriptors such as the research topic, age of the subjects studied, whether the report is a book, journal article, dissertation, etc., and other pertinent information. To conduct the search, the user instructs the computer to find those reports that possess the desired characteristics. For example, you might request the computer to search for articles and books dealing with water and electricity conservation in households in the United States. Note that this search would exclude reports on topics such as conservation of gasoline or natural gas, unpublished studies such as dissertations or technical reports, and studies conducted outside the United States. Carefully defining the extent of the search is necessary to make sure that the user doesn't receive too much or too little information and also to limit the cost of the search.

As noted earlier, research reports are stored as a data base in the computer. A number of different data bases are available, so the researcher must choose which one or ones are most likely to contain the desired information. Data bases exist for psychology, sociology, and other social sciences; and governmental agencies such as the Department of Transportation maintain specialized data bases as well. Normally, researchers and students conduct literature searches with the assistance of a trained librarian who is familiar with the use of a variety of data bases.

A computer literature search is only one example from the field of data base management. Data base management refers to methods of accessing needed information from large data bases. The ability to access information quickly and accurately is, of course, a great advantage of computers, and it is frequently cited as a reason for purchasing a personal computer. For example, consumers are urged to buy a computer so that they can store such things as recipes. Critics of computers point out that an index card could do the job more simply and inexpensively. This is a valid criticism, and many personal computer purchasers have been disappointed to find that a recipe displayed on a computer video monitor is no more informative than the same recipe on a file card.

The criticism would be less valid if the recipes were stored in a data base management system. Here, all items in a list (such as a list of recipes), no matter how large, are coded according to certain characteristics. In the case of recipes, the characteristics may include calories, cost, cooking time, and major ingredients. The user can then decide to access recipes that possess desired characteristics, such as recipes that contain fewer than 600 calories, require less than one hour to cook, and use a particular ingredient that is desired for tonight's dinner.

Used in this way, a computer data base can be a tremendous saver of time and effort and an aid to a fallible human memory.

Computers in the Work Environment

Researchers and students along with people in many professions find that a great deal of their time is spent preparing papers, reports, and correspondence. Computers are changing the way that these tasks are being accomplished. At present, computers are having their greatest impact in the area of word processing.

A word processor is simply a computer equipped with a word processing software program. Many such programs are available for purchase. For example, a salesperson at a local computer store recently showed me a list of 16 different word processing programs that can be purchased for an Apple II™ personal computer. Each program has unique characteristics, and consumers who wish to purchase a computer and/or a word processing program should carefully compare features, including ease of use.

Word processors offer many advantages for both office and personal use. An initial advantage is that typing is done at a computer terminal so corrections are easily made on a screen prior to printing the final product. Also, many word processors offer features such as automatic "wrapping" at the end of each line (thereby eliminating the need to hit a carriage return key), automatic centering, and automatic underlining. Another feature allows the user to insert frequently used material; for example, the word processor in my office is set so that the name of my university, the complimentary close at the end of a letter, and 20 different journal titles can each be typed with just two keystrokes. Some word processors provide for list processing. When form letters are routinely typed, this feature allows the user to type one list of names and addresses and one letter. The letter is then printed over and over with one name and address inserted in the proper place. (List processing is also responsible for the personalized junk mail that most of us receive all too frequently.)

Another advantage is realized when the material is printed. The printed copy can be set with right margin justification or with normal "ragged" right margins.

The major advantage, however, comes when it is time to edit the material. It is easy to add new material at any point, delete as much text as desired, or move a line or paragraph from one place to another. A report or paper doesn't have to be retyped each time it is revised, and in fact word processing encourages people to make revisions.

Word processing is not the only way that computers are changing the work environment. Electronic mail systems are being used to transmit letters and memos from one computer user to another. Bookkeeping, accounting, graph generation, and management spreadsheet programs are extremely popular as well. Because it is easy to link a terminal to a computer via a telephone connection, a great deal of work that previously could only be done at a central office can now be performed at remote locations. Some people are now able to do as much as 50 to 100 percent of their work at home on a computer terminal.

The social impact of such developments has not yet been fully explored. However, many positive results may be cited: time and effort savings, greater flexibility in time scheduling, reduced stress, travel savings, and so on. Negative consequences can also be suggested, including the weakening of interpersonal bonds and productive interactions that can occur in an office environment. There is also concern about the quality of jobs and work satisfaction of people who spend all of their time typing data into a computer. In addition, management has increasing ability to use the computer to monitor work performance and productivity. What might be the psychological effect of such surveillance? Research must be conducted to try to make sure that the positive effects of computer technology are maximized.

Summary

Computers are being used increasingly in education, counseling and clinical psychology, all types of research, and in work settings.

Three types of educational software programs are widely used. The first is traditional computer-assisted instruction (CAI) in which the student is either presented with new material to learn or given drill and practice exercises. A second type of program incorporates games and computer graphics into the exercises. A third type is the use of simulations to model real-life situations. A new development in educational technology incorporates videodiscs into the learning program.

Many educators believe that learning to program a computer is an important educational goal. The reasons for teaching programming are (1) such skills are inherently valuable in a technological society and (2) programming teaches thinking and problem-solving skills. Logo is a programming language designed to achieve these goals with children.

There is concern over computer-related issues such as the quality of educational programs, teacher training, the male-oriented content of educational games, and social class discrepancies in access to computers.

Computerized testing is a new development in counseling and clinical psychology. The computer program administers the test and then prints out scoring and interpretative information. Another development is SIGI, a program designed to provide students with information on their values and career options.

On-line data collection is particularly useful for researchers in field settings. Also, computers may be programmed to control an experiment by means of stimulus presentation, data collection, and statistical analysis.

Archival research data stored by computers provides researchers with an important new tool for investigating research topics in the social sciences.

Computerized literature searches are a valuable aid to students and researchers. Through the search of a computer data base, one can easily obtain a list of the literature on a particular topic.

In the work environment, word processors are replacing typewriters. Word processing greatly simplifies initial typing of information and, most important, eliminates the need for extensive retyping when material must be revised. Electronic mail, computer systems for accounting, and so on are also changing the office environment. The fact that computers allow a great deal of work to be performed at a terminal away from the traditional workplace may have major impacts on the nature of work.

Further Reading

Bork, A. (1981). *Learning with computers*. Maynard, MA: Digital Press.

Barcomb, D. (1981). *Office automation: A survey of tools and techniques*. Maynard, MA: Digital Press.

Behavior research methods and instrumentation. Published proceedings of the National Conference on the Use of On-line Computers in Psychology. (Published annually)

Bulmar, M. (Ed.). (1979). *Censuses, surveys and privacy*. New York: Holmes & Meier.

BYTE. (1982, August). (Logo issue)

College Microcomputer. (Published quarterly)

Hastings, S. (1977, July). Psychiatric assessment via computer. *Creative Computing,* p. 34.

Hyman, H. H. (1972). *Secondary analysis of sample surveys.* New York: Wiley, 1972.

Kiesler, S., Sproull, L., & Eccles, J. S. (1983, March). Second-class citizens? *Psychology Today,* p. 40.

National Opinion Research Center. (1974). *National data program for the social sciences.* Ann Arbor, MI: ICPR.

Papert, S. (1980). *Mindstorm.* New York: Basic Books.

Popular Computing. (1983, August). (Issue on computers in education)

Scientific American. (1982, September). (Issue on the mechanization of work)

Seidel, R. J. (1982). *Computer literacy.* New York: Academic Press.

Taylor, R. P. (Ed.). (1980). *The computer in the school: Tutor, tool, tutee.* New York: Teacher's College Press.

Watt, D. (1983). *Learning with Logo.* New York: BYTE Books/ McGraw-Hill.

Zajonc, R. B. (1975). Birth order and intellectual development. *Psychological Review, 82,* 74–88.

Notes

1. Interestingly, this has led to the creation of programs to unlock the program to allow copying.
2. A similar application of the computer has been found by physicians who use the computer to aid in diagnosis and storing information about individual patients.

Computer
Programming

Computer programs are needed to instruct the computer to perform one or more tasks. For the vast majority of computer users, the actual program is never seen. That is, the program has been written so that the user can input information and receive the desired output while the computer program does all the work. Despite the fact that most of us don't have to write programs directly, it is useful to learn some fundamentals of computer programming in order to understand the logic that underlies the software programs that are available. The goal of this appendix is to provide you with an elementary understanding of programming in BASIC. This will be valuable in increasing your computer literacy. It may also motivate you to learn additional computer skills through further reading and coursework.

Computer Languages

To program a computer, you must be able to speak its language. The fundamental language understood by the computer is called *machine language*. Machine language is difficult and laborious, but ultimately it is the only thing understood by the electronic circuitry of the computer's central processing unit (CPU). To program in machine language, the programmer must tell the computer step by step which memory address is being accessed and provide a direct operation code to add, subtract, and so on. All machine language instructions are written in numbers.

Closely related to machine language is *assembly language*. Assembly language is somewhat simpler than machine language because the programmer can use certain abbreviated letter codes to provide instructions to the computer. Still, assembly language is not something that can be used easily. If you are curious about what assembly language looks like, the following lines of assembly language for the Apple II computer should suffice:

```
CLC
LDA A
ADC B
STA T
LDA A+1
ADC B+1
STA T+1
```

Your immediate reaction might be one of dismay, and then you might question whether it is possible to talk to a computer in some language (such as English!) that is more familiar to you.

Higher Level Languages

Fortunately, computer scientists have developed *higher level languages* that allow programmers to communicate with the computer in a way that is much simpler than either assembly or machine language. Essentially, these languages are used in conjunction with a translator that translates the instructions of the higher order language into instructions that can be directly understood by the computer. Thus, to use a higher level language, the programmer doesn't have to learn how to speak directly to the machine but instead must only learn to speak to a translator who will serve as an intermediary between the programmer and the machine.

A large number of higher level languages are currently in use. Each language has been developed to serve some general purpose of computer use. A brief description of a few of these will provide an overview. The remainder of the appendix will delve more deeply into one higher level language called BASIC.

BASIC

BASIC stands for *B*eginner's *A*ll-Purpose *S*ymbolic *I*nstruction *C*ode. It is a language that is very easy to learn and can be used for a wide variety of applications. Every computer language has certain limitations, however. With BASIC, the programmer has very restricted options in giving instructions to the computer. For example, an instruction such as "multiply the wage per hour times the number of hours worked to determine the amount of pay" must be written in BASIC in a very terse format such as P=W*H.

FORTRAN

FORTRAN is shorthand for *For*mula *Tran*slator. FORTRAN was designed to be used for scientific and mathematical applications of the computer. It is widely used by mathematicians and engineers.

COBOL

COBOL stands for *C*ommon *B*usiness *O*riented *L*anguage. COBOL was developed to provide a language that would be convenient for business applications. Also, COBOL instructions were designed to be closer to spoken English than many other languages.

Conclusion

You will undoubtedly hear references to many other higher level computer languages—Pascal, Logo, APL, C, LISP, and others. If you are curious, the main question to ask is what applications the particular language was designed for. One other point about computer languages must be made. At the present time, there is some variation in how each language is implemented on the various computers available for purchase. That is, each computer company usually has its own version of a certain language that differs in one or more ways from the version of that same language found on the computers of other companies. Thus a program written in BASIC on an IBM (International Business Machines) computer may not work without modification when executed on a CDC (Control Data Corporation) machine. Fortunately, most of the *fundamental* features of BASIC are the same on most machines.

Programming with BASIC

The remainder of this chapter is concerned with the BASIC programming language. As you read this material, a major point to remember is that the only way truly to learn computer techniques is to work actively with a computer. Type in programs, execute them, make mistakes, learn from those mistakes, experiment, and consult with others. Such procedures are invaluable in expanding your abilities.

BASIC programs consist of a series of statements that are instructions to the computer. Each statement is written on a line that contains its own line number. When the computer executes the program, each instruction is carried out sequentially (the first line is executed, then the second, etc.). To illustrate the use of BASIC, consider what we will call Program One:

```
100   REM PROGRAM ONE
110   PRINT "PLEASE TYPE THE NUMBER OF STUDENTS WITH A'S"
120   INPUT A
130   PRINT "PLEASE TYPE THE NUMBER OF STUDENTS WITH B'S"
140   INPUT B
150   PRINT A,B
160   PRINT A;B
170   PRINT
180   PRINT "GOOD-BYE"
190   END
```

If possible, type Program One into your computer. Your instructor will have to provide information on how to log-on to your computer, the initial instruction for creating a new BASIC program file, and the instruction for saving a program file once the program has been input to the computer. To execute a BASIC program, simply type RUN. The

RUN command instructs the computer to execute a BASIC program. To list your program, type LIST. Your program will be shown with line numbers in the proper order. Note that you do not have to type the lines in BASIC in exact sequence from first to last. Lines typed out of order will automatically be inserted in the proper sequence. Also, if you wish to change a line, simply retyping the line will replace the original line with the new one. If you wish to delete a line, typing just the line number and then the return key will remove that line from the program. Let's look at Program One in detail.

First, note that the program begins with line number 100 and lines that follow are in increments of 10. This procedure allows the programmer to enter lines later if that becomes necessary. We will do this as we expand Program One.

The REM Statement

Line 100 contains a REM statement. REM stands for "remark" and such statements are used by the programmer to write comments about the program. In our program, the REM statement was used simply to name the program. REM statements are useful to do such things as write what the program is designed to accomplish and to indicate what particular computer instructions are designed to do. REM statements may be placed anywhere in the computer program; they are not necessary for the program to run but rather are a convenience for the programmer and anyone else who later uses the program. They are particularly useful in long, complex programs. The computer skips over REM statements whenever they appear in a program.

The END Statement

The END statement is *always* the last line of a BASIC program. It tells the computer where to end the program. When the END statement is encountered, no more instructions are processed and the user is able to use the computer for other purposes.

The INPUT Statement

The INPUT statement allows a user to input data or information by typing the information on the terminal. One of the attractive features of BASIC is that programs can be written to be *interactive*. An interactive program prompts the user to provide information. Once the information is input, the computer will do whatever processing is necessary and provide output. The general goal of an interactive program is to make the program "user friendly" (as opposed to "user hostile") so that it is easy to use the program. An INPUT statement appears on line 120. We will discuss the PRINT statement on line 110 in a moment.

When the computer encounters an INPUT statement, the program stops and prints a question mark on the terminal screen. The question mark is a prompt that asks the user to input information. In our program, line 120 would cause a question mark to be printed. What the computer wants is a number that will be referred to as a variable with a particular name. To understand this process, we need to describe the concepts of *variables* and *memory addresses.*

Variables

A piece of information that is to be used by the computer can be assigned to a variable that has a unique variable name. In BASIC, variables are labeled as either single letters, such as A, B, X, N, or as single letter-number combinations, such as A1, A2, B1, B2, X4. Some versions of BASIC allow longer variable names. Usually programmers will attempt to use variable names that are abbreviations for the actual variable. For example, N might be the number of subjects in a study, T might be a total, and so on. It is useful to employ REM statements as reminders of what the variable names refer to.

In line 20, the instruction INPUT A causes the computer to wait for the user to input a number that will represent the value of a variable called A. To make this clearer, suppose that you want the variable A to stand for the number of students who received an "A" grade in a class. The variable named A will be given this value when the program is run. Now suppose that 10 students received "A" grades. When line 120 is executed, the question mark appears, the user types in the number 10 and hits the return button to send the data to the computer. After line 120 is executed, the computer will proceed to line 130 and then 140, INPUT B. It should be obvious that the computer will now produce a question mark asking the user to input the value of the variable named B. Now suppose that B stands for the number of people who received a "B" grade and that 20 students did this. The user would type in the number 20 and hit the return key on the terminal.

Memory Addresses

When information is input to the computer by means of a BASIC program, it is stored in a memory address that has a unique name. In Program One, two memory addresses were created by the INPUT statements on lines 120 and 140. These have the names A and B, and as we executed the program, they took on the values of 10 and 20. You can visualize these memory addresses as follows:

Memory address A currently has the value 10, and address B has the value 20. Whenever the variable name A or B is used in subsequent lines of the program, the computer will refer to the values stored in address A or B and use the values when instructed to do so (for example, a subsequent line might instruct the computer to add the values of A and B together).

The PRINT Statement

In Program One, lines 110, 130, and 150 through 180 each contain a PRINT statement. A PRINT statement directs the computer to print information as output to the user. Recall the first two PRINT statements in Program One on lines 110 and 130:

```
110  PRINT "PLEASE TYPE THE NUMBER OF STUDENTS WITH A'S"
130  PRINT "PLEASE TYPE THE NUMBER OF STUDENTS WITH B'S"
```

Each of these PRINT statements immediately precedes an INPUT statement. An INPUT statement causes a question mark prompt to appear on the screen, and the computer waits for the user to provide input. The purpose of lines 110 and 130 is to inform the user about what sort of information should be input. PRINT statements may be used to print strings and the values of variables.

Printing Strings

In BASIC, a *string* refers to a set of one or more characters. Characters are letters, numbers, punctuation marks, blank spaces, and almost any other symbol on the keyboard. A string appears on both lines 110 and 130. Note that the string is enclosed by quotation marks. The quotation mark must be used; they are not actually printed, but rather they instruct the computer to print whatever is within the quotation marks. When line 110 is executed, the output will simply be

```
PLEASE TYPE THE NUMBER OF STUDENTS WITH A'S
```

Similarly, note that line 180 will give the user a message that the program is over by printing the following string:

```
GOOD-BYE
```

Printing the Values of Variables

Now examine lines 150 and 160:

```
150  PRINT A,B
160  PRINT A;B
```

These PRINT statements direct the computer to print the values of variable A and variable B (more specifically, the contents of memory addresses A and B). There would be two lines of output; the numbers 10 and 20 would be printed on each line.

The only difference in the two PRINT statements is whether the variables are separated by a comma or a semicolon. The use of a comma will cause the values to be printed in columns (usually five columns on a line). The use of a semicolon will cause the values to be printed close together. Thus execution of lines 150 and 160 would cause the following two lines of output:

```
10              20
10  20
```

The use of commas in PRINT statements is convenient when constructing tables. On some versions of BASIC, the semicolon will cause the values to be printed together with *no* space in between. You should experiment with printing variables on your computer.

Printing Blank Lines

Line 170 contains only the PRINT statement with no further instructions. This causes a blank line as output. Blank lines are useful to make the output clearer or more visually appealing. This will be done in most of the programs in this appendix.

Printing Both Strings and Values

It is possible to instruct the computer to print strings and the values of variables on the same line. For example, you could try to modify Program One by substituting lines 150 and 160 with the following two lines:

```
150  PRINT "THE NUMBER OF 'A' GRADES IS "A
160  PRINT "THE NUMBER OF 'B' GRADES IS "B
```

When the program is executed the output will be

```
THE NUMBER OF 'A' GRADES IS 10
THE NUMBER OF 'B' GRADES IS 20
```

Note that if you want to use quotation marks *within* a string, you must use single quotation marks because the only valid use for regular (double) quotation marks is to begin and end a string. Also notice that a blank space was included at the end of the string to create a space in the printed output between the end of the string and the number to be printed.

PRINT statements help make the program more interactive and "user friendly." They can be used for greetings, instructions, prompts, or congratulatory messages to the user. Take note of some of the ways in which PRINT statements are used in subsequent programs in this appendix.

String Variables

Strings may also be used as variables. That is, a string of characters may be stored in a unique memory address and used in a program. String variable names consist of a letter and a dollar sign—A$, B$, N$, and so on. Program Two creates and uses a string variable, N$.

```
100  REM PROGRAM TWO
110  PRINT "HELLO THERE."
120  PRINT "I'M YOUR FRIENDLY COMPUTER."
130  PRINT "WHAT'S YOUR NAME?"
140  PRINT
150  INPUT N$
160  PRINT
170  PRINT "GLAD TO MEET YOU " N$
180  PRINT "I'M HAPPY TO BE OF SERVICE."
190  PRINT "GOOD-BYE FOR NOW, " N$
200  END
```

The functions of lines 100 through 140 should be familiar to you. Line 150 is an INPUT statement, but instead of asking for a numeric variable, a string variable is requested. Note that the INPUT statement was preceded by a printed request to type in a name. It is always good practice to provide the user with an instruction preceding the INPUT statement that will in effect explain what is wanted when the question mark prompt appears.

When Program Two is run by someone named Sue, the output looks like this:

```
HELLO THERE.
I'M YOUR FRIENDLY COMPUTER.
WHAT'S YOUR NAME?

?SUE
```

GLAD TO MEET YOU SUE
I'M HAPPY TO BE OF SERVICE.
GOOD-BYE FOR NOW, SUE

Note that the user did not have to use quotation marks when typing her name. When she did input her name, the string of characters was stored in a memory address labeled N$. You can visualize this as

N$

Whenever the string variable, N$, is used later in the program, the computer will use the contents of this memory address. Thus, in lines 170 and 190, the computer is instructed to print the SUE that is stored as the string variable called N$.

Arithmetic Operations

One of the most frequent uses of a BASIC program is to perform calculations on numbers stored in memory. The following symbols are used in BASIC:

+ Adds numbers
− Subtracts numbers
* Multiplies numbers
/ Divides numbers
∧ Exponentiates numbers

To illustrate, recall that in Program One we stored the number 10 as the value of A and 20 as the value of B. We might now add lines that perform arithmetic operations and output the results. Consider the following examples as "inserts" into Program One:

Line in Program	Output
161 PRINT A+B	30
162 PRINT B−A	10
163 PRINT A*B	200
164 PRINT B/A	2
165 PRINT A ∧ 2	100
166 PRINT A ∧ 3	1000

The statement PRINT A+B causes 10 to be added to 20 and the result (30) is output. Remember that exponentiation is simply multiplying a number by itself. Thus, A ∧ 2 is 10 × 10 and A ∧ 3 is 10 × 10 × 10. The other examples should be self-explanatory.

When more than one arithmetic operation is carried out in a single statement, it is important to carefully write the instruction so that the computer does the correct calculations. To illustrate the problem, ask yourself whether A+B/A+B should be 30/30 = 1 or 10 + 2 + 20 = 32. There is an ambiguity in the equation A+B/A+B regarding whether the addition or the division should be carried out first. BASIC has definite rules for deciding which operations have priority. The rules may be summarized as follows:

Rule 1: ∧ has highest priority. Exponentiation will be carried out first.

Rule 2: / and * have next highest priority.

Rule 3: + and − have lowest priority.

Rule 4: When two or more operations of the same priority appear in an instruction, they will be carried out in the order in which they appear (from left to right).

Rule 5: Instructions within parentheses are carried out as a separate unit within the equation. These are carried out before all other calculations.

The following are several examples of applying these rules:

A+B+A ∧ 2 = 10 + 20 + 100 = 130   (Rule 1)

A+B*B/A−B = 10 + 400/10 − 20 = 30   (Rules 2, 3, 4)

A*(B+A)+B = 10 × 30 + 20 = 300 + 20 = 320   (Rules 2, 3, 5)

If you write your own programs, it is extremely important to check such instructions for accuracy. You must determine whether the computer's calculation produces the correct answer. When writing the statements to perform calculations, it is often desirable to break the mathematical equation down into two or more parts. For example, (A+B)/(C+D) could be calculated with three lines:

X = A + B
Y = C + D
Z = X/Y

This is done in BASIC by means of LET statements.

The LET Statement

We have already used the INPUT statement to create numeric or string variables and store values for the variables in the computer's memory. Another way of creating variables and storing information is by means of the LET statement. Here are two examples:

```
100   LET N=100
110   LET N$="SUE NORTH"
```

In line 100, a numeric variable called N was created with the value 100 stored in memory. In line 110, a string variable was created; N$ takes on the value SUE NORTH. Notice that when we refer to a string variable within the program, quotation marks must be used to enclose the string.

In actuality, the word LET is not necessary on most computers. The following lines are understood by the computer as identical to lines 100 and 110 above.

```
100   N=100
110   N$="SUE NORTH"
```

The main reason for omitting the word LET from these statements is to save time when writing and typing the program.

It is possible to use other variables in a LET statement. For example, in Program One, the following line would create a new variable that is the *total* number of grades:

```
167 LET T=A+B
```

The effect of line 167 is to add together the values currently stored in A and B, and to store this total in a memory location T. We now have three memory addresses:

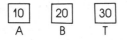

It would now be possible to expand the program to do such things as print the total or calculate the percentage of students who received grades of "A" and "B."

It is also possible to use LET statement expressions to replace the contents of a memory address with a new value. The ability to do this is extremely useful in programs that require a constantly changing value such as a counter or to calculate a total (later programs will do this). To illustrate, consider the effects of the following two lines:

```
100  LET A=10
110  LET A=20
```

When the computer executes line 100, a memory location A is given the value 10. However, when the next line is executed, the computer must go to memory location A and replace the current value (10) with the new value (20). There cannot be two memory locations with the same name.

A more complex example is a counter. Consider the following three lines:

```
100  LET N=0
110  LET N=N+1
120  LET N=N+1
```

How would the computer execute these three lines? Line 100 would create a memory address N with the value of 0.

N

Line 110 would perform the calculation N=N+1; because the current value of N is zero, this would be calculated as N = 0 + 1. The result would be stored in memory location N and replace the old value of N. Thus our memory address is now

N

When line 130 is executed, the expression N=N+1 is calculated as N = 1 + 1 because the value of 1 is the current value of N. The result of the calculation becomes the new value of N:

N

As we will see shortly, it is possible to write useful programs that do such counting very easily.

The GOTO Statement
A GOTO statement directs the computer to immediately *go to* a specified line number and begin executing at that point. The line the computer

goes to may precede or follow the GOTO statement. Program Three uses a GOTO statement to produce a continual loop. Programmers strive to avoid loops that won't allow the user to easily exit from the program; however, Program Three is always good as a self-esteem booster.

```
100   REM PROGRAM THREE
110   PRINT "PLEASE TYPE YOUR NAME."
120   PRINT
130   INPUT N$
140   PRINT "YOU ARE WONDERFUL, "N$
150   GOTO 140
160   END
```

People using this program would find the computer telling them that they are wonderful over and over and over again. The reason is that the GOTO command forces the computer to execute over and over the instructions on line 140.

The IF THEN Statement

A major problem of the simple GOTO statement is that it often results in undesirable loops that can't be stopped. Much more useful are IF THEN statements that take the form, *if* something is true *then* proceed to another part of the program. This is written in a form such as

```
100   IF A=B THEN 200
```

In this example, when the computer executes line 100, a comparison will be made: If the value of A is equal to the value of B, the computer will skip to line 200. If the value of A is not equal to the value of B, the computer will continue on and execute the next line in the program. There are six types of comparisons that can be made in an IF THEN statement:

IF A=B	Is A equal to B?
IF A>B	Is A greater than B?
IF A<B	Is A less than B?
IF A>=B	Is A greater than or equal to B?
IF A<=B	Is A less than or equal to B?
IF A<>B	Is A not equal to B?

A modified version of Program Three will serve as the first illustration of the IF THEN statement. This program also uses a counter variable, N, that we set to zero at the beginning of the program.

```
100  REM PROGRAM THREE REVISED
110  LET N=0
120  PRINT "PLEASE TYPE YOUR NAME."
130  PRINT
130  INPUT N$
140  PRINT "YOU ARE WONDERFUL, "N$
150  LET N=N+1
160  IF N<50 THEN 140
170  END
```

In this program, line 150 sets up a counter: every time YOU ARE WONDERFUL is printed, the value of N advances by one. Now examine the IF THEN statement on line 160. As long as the value of N is less than 50, the computer will continue to return to line 140 and print the output. However, as soon as the value of N reaches 50 (that is, YOU ARE WONDERFUL has been printed 50 times), the computer will no longer go to line 140 but will instead proceed to the next line of the program. (*Note:* The next section on FOR-NEXT loops will perform the same function more efficiently.)

Program Four illustrates a more practical use of the IF THEN statement. This program calculates the amount of someone's pay using different formulas depending on whether the person worked more than 40 hours during the week.

```
100  REM *************************
110  REM PROGRAM FOUR
120  REM A WAGE PROGRAM WRITTEN BY
130  REM SAM SPADE
140  REM *************************
150  PRINT
160  PRINT "HOW MANY HOURS WERE WORKED"
170  INPUT H
180  PRINT "WHAT IS THE WAGE PER HOUR"
190  INPUT W
200  IF H>40 THEN 230
210  LET P=H*W
220  GOTO 240
230  LET P=(40*W)+(H-40)*(W*1.5)
240  PRINT
250  PRINT "THE PAY IS "P
490  PRINT "GOOD-BYE"
500  END
```

There are several things to point out about Program Four. The first thing you may have noticed is the five lines of REM statements. These were included to show you a convenient method of labeling your program and identifying it as written by you. More important, the IF THEN statement appears at line 200. If the number of hours worked is greater than 40, the computer will proceed to line 230, use the overtime formula (hours over 40 are paid at time-and-a-half wages) to calculate P, and then print the amount of pay. However, if the number of hours is not greater than 40, the computer will simply proceed to the next line and use the simple formula for calculating pay (P=H*W). If the computer takes this route, a GOTO statement follows to direct the computer to go to line 240 to print the amount of pay.

You may have noticed that the END statement was given line number 500. In fact, it is frequently a good idea to start out writing a program by typing the END statement with a very large line number. However, in Program Four the purpose was to provide room to modify the program to allow the user to perform another calculation. The following lines inserted into Program Four would perform this function.

```
260  PRINT
270  PRINT "DO YOU HAVE ANOTHER PROBLEM"
280  INPUT A$
290  IF A$="YES" or A$="Y" THEN 150
```

After the first problem is calculated, the program asks if the user has another problem. The INPUT statement will produce a question mark for the answer (A$). An IF THEN statement comes next. *If* the user types in either YES or simply Y, *then* the computer returns to the beginning of the program at line 150. If the user types anything else, the computer proceeds to the next line and eventually ends the program. This makes the program more interactive and user friendly. (*Note:* A slightly more sophisticated method of looping for additional runs is illustrated later in a program called Beerhunter.)

You might note that the word OR was used in the IF THEN statement on line 290. This demonstrates that the IF conditions can be more complex. For example, they can take either of the following forms:

```
100  IF A=B OR A=C THEN 200
100  IF A=B AND A=C THEN 200
```

In the first case, the computer would go to line 200 whenever the value of A is equal to either B *or* C. In the second case, it would proceed to line 200 only when A is equal to the value of both B *and* C.

The FOR-NEXT Loop

The FOR and NEXT statements (usually called a FOR-NEXT loop) serve to build into the program a counter that will execute BASIC commands within the loop as many times as desired. Consider Program Five:

```
100   REM PROGRAM FIVE
110   FOR I=1 TO 10
120   PRINT "HELLO", I
130   NEXT I
140   END
```

The FOR statement on line 110 serves to specify the number of loops; the I (letter I) variable serves as a counter. The NEXT statement must always be used at the end of the loop. Line 110 specifies that the program will execute the loop ten times; whatever commands are placed within the loop will be executed that many times. After each loop, the counter will increment the value of variable I and a check will be made to see whether ten loops have been executed. After looping the designated number of times, the computer proceeds to the line that follows the NEXT statement.

What will the output of this program look like? The only statement within the loop is a command to print the word HELLO and then print the current value of variable I. Because the value of I changes with each execution of the loop, the output will be

```
HELLO                   1
HELLO                   2
HELLO                   3
HELLO                   4
HELLO                   5
HELLO                   6
HELLO                   7
HELLO                   8
HELLO                   9
HELLO                  10
```

Frequently, it is desirable to allow the user to determine the number of loops to be executed. This can be accomplished with an INPUT statement that asks the user to input a number that will be used in the FOR statement. Program Six is a whimsical example of this.

```
100   REM PROGRAM SIX
110   PRINT "HOW DO I LOVE THEE?"
120   PRINT "LET ME COUNT THE WAYS."
130   PRINT "PLEASE INPUT THE NUMBER OF WAYS."
140   INPUT N
150   FOR I=1 TO N
160   PRINT "I LOVE THEE"
170   NEXT I
180   END
```

When the user inputs a number, it becomes the value of the variable N. The FOR statement is written so that the loop will be executed N number of times. If the user inputs 2, I LOVE THEE will be printed twice. If the user inputs 1000, the printed output will appear a thousand times.

A more serious application of this technique is provided by Program Seven. This program is designed to calculate the sum of a set of scores and then calculate the mean (average) of those scores. The statistical formula for the sum of the scores is ΣX (where Σ is the symbol for a sum and X is the symbol for a score). The formula for the mean is $\Sigma X/N$ (where N is the number of scores).

```
100   REM ******************************
110   REM PROGRAM SEVEN
120   REM SUM AND MEAN OF A SET OF SCORES
130   REM ******************************
140   REM SET SUM VARIABLES TO ZERO
150   LET S=0
160   PRINT
170   PRINT "HOW MANY SCORES?"
170   INPUT N
180   PRINT
190   PRINT "PLEASE INPUT YOUR SCORES ONE AT A TIME"
200   PRINT
210   REM LOOP TO INPUT AND SUM SCORES
220   FOR I=1 TO N
230   INPUT X
240   REM X IS THE SCORE AND S IS THE SUM
250   LET S=S+X
260   NEXT I
270   PRINT
280   PRINT "THE SUM = "S
290   REM M IS THE MEAN
300   LET M=S/N
310   PRINT
320   PRINT "THE MEAN = "M
330   PRINT
340   PRINT "DO YOU HAVE ANOTHER PROBLEM?"
350   INPUT A$
360   IF A$="YES" OR A$="Y" THEN 150
370   PRINT "GOOD-BYE FOR NOW"
380   END
```

Note that this program incorporates a number of features of BASIC programming that have been presented earlier (REM, PRINT, allowing multiple problems). To understand more fully how this program works, we can examine the contents of the various memory locations as a user inputs three scores: 1, 2, and 3. Notice that this program has the variables S, N, I, X, and M and that only the variables in the loop change once set:

Line	S	N	I	X	M	Operation
Line 150	0					S set to zero
Line 170	0	3				Input N
Line 220	0	3	1			Loop begins
Line 230	0	3	1	1		Input X
Line 250	1	3	1	1		SUM=S+X
Line 260	1	3	2	1		Loop to 230
Line 230	1	3	2	2		Input X
Line 250	3	3	2	2		SUM=S+X
Line 260	3	3	3	2		Loop to 230
Line 230	3	3	3	3		Input X
Line 250	6	3	3	3		SUM=S+X
Line 260						Loop ends
Line 300	6	3	3	3	2	M=S/N

You might wonder why the value of S was set to zero on line 150. The reason is that if the user chooses to input another set of scores, the value of S would not have changed when the program gets to line 230. Thus it is good practice to set variables that change within the loop to zero. Variables outside the loop such as M or N do not have to be set to zero at the beginning of the program.

It would be very easy to expand Program Seven to perform additional calculations either within the loop or outside it. For example, many statistical formulas require finding the sum of the squared value of each score, ΣX^2 and/or the squared value of the sum, (ΣX^2). These could be calculated with the following three lines:

255 LET S2=S2+X $\wedge$ 2	(*Note:* This is inserted within the loop.)
285 LET S3=S $\wedge$ 2	(*Note:* This is inserted outside the loop.)
155 LET S2=0	(*Note:* Set variable that changes within loop to zero.)

Built-In Functions

BASIC contains a number of built-in functions that can be used to automatically perform certain operations. One of the most useful is SQR(X), which calculates the square root of the value of any variable within the parentheses. If it were necessary to find the square root of the sum in Program Seven, the following line would do it:

```
286  LET Q = SQR(S)
```

The new variable Q contains the value of the square root of S. Other built-in functions include

ABS(X)   Absolute value of X
INT(X)   Make X a whole integer
RND   Provide a random number between 0 and 1

Subroutines

The programs described thus far have progressed in a line-by-line fashion; that is, each line is executed in sequence from first to last when the program is run. A problem with this type of program, particularly when it is a complex one, is that it is difficult for a person who did not write the program to clearly read, debug, or modify it. To be able to do this, programmers are taught to break the program into self-contained modules. Each module contains the computer instructions necessary to perform a particular part of a complex computer program. An attempt is made to *structure* the program so that it is constructed of a set of modules. One criticism of the BASIC programming language is that it does not easily allow such structured programming. There are several BASIC statements, however, that help a programmer use modules. We will focus on one statement that creates subroutines. A *subroutine* is a self-contained set of lines within the program that performs a particular function within the program. The statement is called GOSUB.

A GOSUB statement takes the program to a particular line to execute a subroutine. The subroutine ends with a RETURN instruction that takes the program back to the line that immediately follows the original GOSUB statement. Examine Program Eight, which uses a subroutine.

```
100   REM **********************************************
110   REM PROGRAM EIGHT
120   REM **********************************************
130   LET S=0
140   PRINT "HOW MANY SCORES"
```

```
150  INPUT N
160  GOSUB 230
170  PRINT
180  GOSUB 320
190  PRINT "DO YOU HAVE ANOTHER PROBLEM"
200  INPUT A$
210  IF A$="YES" or A$="Y" THEN 130
220  GOTO 370
230  REM *********************************************
240  REM SUBROUTINE FOR INPUT OF SCORES
250  REM *********************************************
260  PRINT "PLEASE TYPE IN THE SCORES"
270  FOR I=1 TO N
280  INPUT X
290  LET S=S+X
300  NEXT I
310  RETURN
320  REM *********************************************
330  REM SUBROUTINE FOR CALCULATING STATISTICS
340  LET M=S/N
350  PRINT "THE MEAN = "M
360  RETURN
370  PRINT "GOOD-BYE"
380  END
```

Program Eight performs the same job as Program Seven. However, subroutines have been used in Program Eight to clearly demarcate the portions of the program that require input of the scores and the calculation of the statistics. In a more complex program, the subroutines would contain many more statements. You might now notice that because the subroutine is a small program within the main program, any modifications to the subroutine do not affect the whole program. This is a great advantage to a programmer who is attempting to debug or modify a program.

A Simulation Program

A game program will be used to conclude this chapter and demonstrate many of the BASIC principles discussed so far. You are probably familiar with the game of Russian roulette. Bob and Doug McKenzie of "Second City Television," a show that was on NBC, have come up with their less violent version of Russian roulette, a game called The Beerhunter. The game is played with a six-pack of beer; one beer is vigorously shaken and mixed with the others. Then the player puts the can to his head and opens the beer. If the beer explodes all over his head, the player loses and is a "wethead."

We can write a program to simulate the Beerhunter game. This program will have to do the following:

1. Print instructions.
2. Request whether the user wishes to play or exit from the program.
3. Use a random number to determine whether the player has won or lost.
4. Print appropriate messages, depending on a win or a loss.
5. Have a counter that keeps track of the number of wins.
6. Congratulate the player when he or she has won five times, and end the program.

To obtain a random number, the RND function can be used. Recall that RND produces a random number between 0 and 1. Because there is one chance out of six that the shaken beer will be picked, our program can state that one sixth of the possible random numbers will constitute a "loss," whereas five sixths of the numbers will be a "win." In the program, any random number between 0 and .16 will be a loss; any number between .17 and .99 will be a win.

Figure 1 shows the program. You can type this program in your computer and play the game. One note of caution concerns the randomization function. Each computer requires the user to give slightly different instructions to use random numbers. Most machines require the line with the word RANDOMIZE; this is done to begin the program with a new set of random numbers each time it is run. However, some computers have other methods of starting a sequence of random numbers. Consult your instructor or the manual of your computer to find out exactly how randomization operates on your computer.

Conclusion

The major objective of this appendix has been to familiarize you with the procedures and logic of computer programming. The appendix has provided you with a great deal of information about BASIC programming language. There are a number of other features and capabilities of BASIC, but they are beyond the scope of this book. If your curiosity and interest have been aroused, you may wish to pursue further coursework in BASIC or other computer languages.

Figure 1   The Beerhunter Program

```
100 REM **THE BEERHUNTER**
110 REM ** A GAME INSPIRED BY THE GREAT WHITE NORTH**
120 RANDOMIZE
130 PRINT
140 PRINT "THE BEERHUNTER"
150 PRINT
160 PRINT "HI, I'M BOB MCKENZIE AND I'M HERE WITH MY HOSEHEAD BROTHER, DOUG."
170 PRINT "GOOD DAY"
180 PRINT "WE'RE HERE TO HELP YOU PLAY BEERHUNTER."
190 PRINT "WE'VE GOT SIX CANS OF BEER ON THE TABLE HERE AND WE'RE SHAKING"
200 PRINT "ONE UP REAL GOOD."
210 PRINT "IN BEERHUNTER, YOU PICK OUT A CAN, HOLD IT TO YOUR HEAD"
220 PRINT "AND POP IT OPEN. IT'S KIND OF LIKE RUSSIAN ROULETTE, EH?"
230 PRINT "IF YOU LOSE, YOU'RE A WETHEAD, BUT IF YOU WIN"
240 PRINT "YOU GET TO DRINK THE BEER."
250 LET N=0
260 PRINT
270 PRINT "SO DO YOU WANT TO PLAY BEERHUNTER, EH?"
280 PRINT
290 INPUT A$
300 IF A$="YES" OR A$="Y" THEN 340
310 IF A$="NO" OR A$="N" THEN 510
320 PRINT "PLEASE TYPE 'YES' OR 'NO' "
330 GOTO 290
340 IF RND < .17 THEN 420
350 LET N=N+1
360 PRINT
370 PRINT "POP"
380 PRINT "YOU WIN THIS TIME . . . DRINK UP."
390 IF N=5 THEN 460
400 PRINT "TRY AGAIN?"
410 GOTO 280
420 PRINT
430 PRINT "SPLASH !!!!!"
440 PRINT "YOU LOSE, WETHEAD."
450 GOTO 250
460 PRINT
470 PRINT "YOU'VE ALMOST DRUNK A WHOLE SIXPACK NOW"
480 PRINT "AND YOU'RE REALLY HOSED. CONGRATULATIONS."
490 PRINT "YOU WON BEERHUNTER."
500 GOTO 520
510 PRINT "TAKEOFF, HOSER"
520 PRINT "GOOD DAY"
530 END
```

Note: This program does not play a true game of Russian roulette because the probability of losing does not change each time there is a win (i.e., the probability changes from one-sixth to one-fifth to one-fourth to one-third to one-half). Could you alter the program so that the probability of losing would change each time?

Summary

All functions carried out by a computer are controlled by instructions in machine language. Programming in machine language is very difficult to do. Higher level languages such as BASIC allow programming in a simpler language that the computer eventually translates into machine language. BASIC stands for *B*eginner's *A*ll-Purpose *S*ymbolic *I*nstruction *C*ode.

A BASIC program consists of a series of lines; each line contains a line number and a statement that gives the computer an instruction. When the program is executed, lines are executed in sequence. RUN is the command to execute the program, and LIST is used to list the contents of the program. The last line in the program must be END.

Variables are pieces of information stored by the computer in a particular memory address. Numeric variables contain numbers and have variable names consisting of letters or letters plus a number (e.g., A, B, A1, A2). String variables consist of a string of characters and have variable names consisting of a letter and dollar sign (e.g., A$, N$).

The REM statement allows the programmer to make a remark. It is not processed by the computer. The PRINT statement is used to (1) print a blank line, (2) print a string of characters within quotation marks, or (3) print the value of a variable.

The INPUT statement allows the user to input the value of a variable. The LET statement assigns a value to a variable within the program (e.g., LET X=10). LET statements are also used to perform arithmetic operations (e.g., LET A=X+Y).

The GOTO statement instructs the computer to go to a specified line in the program and begin executing at that point.

The IF THEN statement instructs the computer to go to a specified line in the program only if a certain condition is true.

The FOR and NEXT statements are used to create a FOR-NEXT loop. Instructions within the loop are executed a specified number of times.

Subroutines are self-contained sets of lines that perform a particular function. The GOSUB statement instructs the computer to go to a certain line that begins the subroutine. The subroutine ends with RETURN, and the computer proceeds to the line following the initial GOSUB statement. Subroutines help to structure the program.

Further Reading

Graham, N. *The mind tool.* (1983). St. Paul, MN: West.

Hennefield, J. (1981) *Using BASIC.* Boston: Prindle, Weber & Schmidt.

Wolach, A. H. (1983). *Basic analysis of variance programs for micro-computers.* Monterey, CA: Brooks/Cole.

Index